T0141982

Advances in Intelligent Systems and Computing

Volume 372

About this Series

The series "Advances in Intelligent Systems and Computing" contains publications on theory, applications, and design methods of Intelligent Systems and Intelligent Computing. Virtually all disciplines such as engineering, natural sciences, computer and information science, ICT, economics, business, e-commerce, environment, healthcare, life science are covered. The list of topics spans all the areas of modern intelligent systems and computing.

The publications within "Advances in Intelligent Systems and Computing" are primarily textbooks and proceedings of important conferences, symposia and congresses. They cover significant recent developments in the field, both of a foundational and applicable character. An important characteristic feature of the series is the short publication time and world-wide distribution. This permits a rapid and broad dissemination of research results.

More information about this series at http://www.springer.com/series/11156

Javier Bajo · Josefa Z. Hernández
Philippe Mathieu · Andrew Campbell
Antonio Fernández-Caballero · María N. Moreno
Vicente Julián · Amparo Alonso-Betanzos
María Dolores Jiménez-López · Vicente Botti
Editors

Trends in Practical Applications of Agents, Multi-Agent Systems and Sustainability

The PAAMS Collection

Editors
Javier Bajo
Departamento de Inteligencia Artificial
Universidad Politécnica de Madrid
Madrid, Spain

Josefa Z. Hernández
Departamento de Inteligencia Artificial
Universidad Politécnica de Madrid
Madrid, Spain

Philippe Mathieu
Lille University of Science and Technology
Villeneuve d'Ascq Cédex, France

Andrew Campbell
Department of Computer Science
Dartmouth College
Hanover, USA

Antonio Fernández-Caballero
Universidad de Castilla-La Mancha ESII
Computing Systems Department
Albacete, Spain

María N. Moreno
Departamento de Informática y Automática
Universidad de Salamanca
Salamanca, Spain

Vicente Julián
Departamento de Sistemas Informáticos y Computación
Universidad Politécnica de Valencia
Valencia, Spain

Amparo Alonso-Betanzos
Laboratory for Research and Development in Artificial Intelligence (LIDIA)
Computer Science Department
University of A Coruña
A Coruña, Spain

María Dolores Jiménez-López
Facultat de Lletres de la Universitat Rovira i Virgili
Tarragona, Spain

Vicente Botti
Departamento de Sistemas Informáticos y Computación
Universidad Politécnica de Valencia
Valencia, Spain

ISSN 2194-5357 ISSN 2194-5365 (electronic)
Advances in Intelligent Systems and Computing
ISBN 978-3-319-19628-2 ISBN 978-3-319-19629-9 (eBook)
DOI 10.1007/978-3-319-19629-9

Library of Congress Control Number: 2015940026

Springer Cham Heidelberg New York Dordrecht London

Printed on acid-free paper

Springer International Publishing AG Switzerland is part of Springer Science+Business Media (www.springer.com)

Preface

PAAMS'15 Special Sessions are a very useful tool in order to complement the regular program with new or emerging topics of particular interest to the participating community. Special Sessions that emphasized on multi-disciplinary and transversal aspects, as well as cutting-edge topics were especially encouraged and welcome.

Research on Agents and Multi-Agent Systems has matured during the last decade and many effective applications of this technology are now deployed. An international forum to present and discuss the latest scientific developments and their effective applications, to assess the impact of the approach, and to facilitate technology transfer, has become a necessity.

PAAMS, the International Conference on Practical Applications of Agents and Multi-Agent Systems is an evolution of the International Workshop on Practical Applications of Agents and Multi-Agent Systems. PAAMS is an international yearly tribune to present, to discuss, and to disseminate the latest developments and the most important outcomes related to real-world applications. It provides a unique opportunity to bring multi-disciplinary experts, academics and practitioners together to exchange their experience in the development of Agents and Multi-Agent Systems.

This volume presents the papers that have been accepted for the 2015 special sessions: Agents Behaviours and Artificial Markets (ABAM); Agents and Mobile Devices (AM); Multi-Agent Systems and Ambient Intelligence (MASMAI); Web Mining and Recommender systems (WebMiRes); Learning, Agents and Formal Languages (LAFLang); Agent-based Modeling of Sustainable Behavior and Green Economies (AMSBGE); Emotional Software Agents (SSESA) and Intelligent Educational Systems (SSIES). The volume also includes the paper accepted for the Doctoral Consortium in PAAMS 2015.

We would like to thank all the contributing authors, the members of the Program Committee, the sponsors (IEEE SMC Spain, IBM, AEPIA, AFIA, AAAI, APPIA, ARIA, ATIA, BNVKI, SADIO, SBC, KI, University of Salamanca and CNRS) and the Organizing Committee for their hard and highly valuable work. Their work has helped to contribute to the success of the PAAMS'15 event. Thanks for your help – PAAMS'15 would not exist without your contribution.

Javier Bajo
PAAMS'15 Organizing Committee Chair

Organization

Special Sessions

Agents Behaviours and Artificial Markets
Agents and Mobile Devices
Learning, Agents and Formal Languages (LAFLang)
Multi-Agent Systems and Ambient Intelligence
Web Mining and Recommender systems
Agent-based Modeling of Sustainable Behavior and Green Economies
Intelligent Educational Systems
Emotional Software Agents

Special Session on Agents Behaviours and Artificial Markets

Philippe Mathieu (Chair)	University of Lille, France

Scientific Committee

Bruno Beaufils	University of Lille1, France
Olivier Brandouy	University of Paris 1, France
Florian Hauser	Austria, Austria
Philippe Mathieu	University of Lille1, France
Adolfo López Paredes	University of Valladolid, Spain
Javier Arroyo	University Complutense Madrid, Spain
Marco Raberto	University of Genoa, Italy
Roger Waldeck	Telecom Bretagne, France
Murat Yildizoglu	University of Bordeaux IV, France

Special Session on Agents and Mobile Devices

Andrew Campbell	Darthmouth College, USA
Javier Bajo	Polytechnic University of Madrid, Spain

Scientific Committee

Antonio Juan Sánchez	University of Salamanca, Spain
Juan Francisco De Paz	University of Salamanca, Spain
Gabriel Villarrubia	University of Salamanca, Spain
Cristian Pinzón	Technical University of Panama, Panama
Montserrat Mateos	Pontifical University of Salamanca, Spain
Luis Fernando Castillo	University of Caldas, Colombia
Miguel Ángel Sánchez	Indra, Spain
Roberto Berjón	Pontifical University of Salamanca, Spain
Encarnación Beato	Pontifical University of Salamanca, Spain
Fernando De la Prieta	University of Salamanca, Spain

Special Session on Learning, Agents and Formal Languages (LAFLang)

María Dolores Jiménez López	Universitat Rovira i Virgili, Spain
Leonor Becerra-Bonache	University of Saint-Etienne, France

Special Session on Multi-Agent Systems and Ambient Intelligence

Antonio Fernández-Caballero (Co-Chair)	Uni. of Castilla-La Mancha, Spain
Elena María Navarro (Co-Chair)	University of Castilla-La Mancha, Spain

Scientific Comittee

Javier Jaén-Martínez	Universidad Politécnica de Valencia, Spain
Mª Teresa López-Bona	University of Castilla-La Mancha, Spain
Diego López-de-Ipiña	University of Deusto, Spain
Rafael Martínez-Tomás	National Distance Education University, Spain
Paulo Novais	Universidade do Minho, Spain
Juan Pavón	University Complutense Madrid, Spain

Special Session on Web Mining and Recommender Systems

María Moreno (Chair)	University of Salamanca, Spain

Scientific Committee

Ana María Almeida	Institute of Engineering of Porto, Portugal
Yolanda Blanco Fernández	University of Vigo, Spain
Rafael Corchuelo	University of Sevilla, Spain
Chris Cornelis	Ghent University, Belgium
María José del Jesús Díaz	University of Jaén, Spain
Anne Laurent	University of Montpellier 2, France
Vivian López Batista	University of Salamanca, Spain
Joel Pinho Lucas	Mobjoy, Brazil
Constantino Martins	Institute of Engineering of Porto, Portugal
Yong Zheng	DePaul University, USA

Special Session on Agent-Based Modeling of Sustainable Behavior and Green Economies

Amparo Alonso-Betanzos	University of A Coruña, Spain
Noelia Sánchez-Maroño	University of A Coruña, Spain
Óscar Fontenla-Romero	University of A Coruña, Spain
Gary Polhill	The James Hutton Institute, UK
Tony Craig.	The James Hutton Institute, UK

Special Session on Intelligent Educational Systems

Juan Miguel Alberola (Co-Chair)	Universitat Politecnica de Valencia, Spain
Elena Del Val (Co-Chair)	Universitat Politecnica de Valencia, Spain
Vicente Julian (Co-Chair)	Universitat Politecnica de Valencia, Spain

Special Session on Emotional Software Agents

Vicente Botti	Universitat Politècnica de València, Spain
Vicente Julián	Universitat Politècnica de València, Spain
Emilio Vivancos	Universitat Politècnica de València, Spain
Bexy Alfonso	Universitat Politècnica de València, Spain

Scientific Committee

Sara Rodríguez	University of Salamanca, Spain
Eugénio Oliveira	University of Porto, Portugal
Javier Bajo	Polytechnic University of Madrid, Spain
John-Jules Ch. Meyer	University Utrecht, The Netherlands
Juan F. De Paz	University of Salamanca, Spain
Catherine Pelachaud	CNRS at TELECOM ParisTech, France
Alberto Fernández	University Rey Juan Carlos, Spain
Eva Hudlicka	University of Massachusetts Amherst, USA
Sascha Ossowski	University of Rey Juan Carlos, Spain
Stefan Rank	Drexel University in Philadephia, USA
Cesar Analide	University of Minho, Portugal
Holger Billhardt	Universidad Rey Juan Carlos, Spain
Carlos Carrascosa	Universidad Politécnica de Valencia, Spain

PAAMS 2015 Special Sessions Organizing Committee

Javier Bajo (Chair)	Technical University of Madrid, Spain
Fernando de la Prieta Pintado	University of Salamanca, Spain
Juan F. De Paz	University of Salamanca, Spain
Sara Rodríguez	University of Salamanca, Spain
Gabriel Villarrubia González	University of Salamanca, Spain
Javier Prieto Tejedor	University of Salamanca, Spain
Pablo Chamoso	University of Salamanca, Spain
Alberto López Barriuso	University of Salamanca, Spain

PAAMS 2015 Sponsors

Contents

Part I

Special Session on Agents Behaviours and Artificial Markets (ABAM)

Behavioral and Informational Agents-Based Modeling and Simulation of Emerging Stock Markets

Hazem Krichene[1] and Mhamed-Ali El-Aroui[1,2]

[1] LARODEC, ISG de Tunis, Tunisia
krichene.hazem@gmail.com
[2] Université de Carthage, FSEG Nabeul, Tunisia
mhamed.elaroui@yahoo.com

Abstract This work presents an artificial order-driven market populated by heterogeneous agents characterized by mixed behaviors and shared information. This market is designed to reproduce immature markets stylized facts based mainly on important asymmetric information and herd behaviors. Information flows are modeled by a directed weighted network. The important information asymmetry is modeled by different assortative network behaviors. Our experimental findings show that the artificial market developed here was able to reproduce the main features characterizing immature stock markets.

Keywords: Agent-Based Model, Immature Financial Markets, Network Theory, Information Asymmetry, Herd Behavior, Assortative Network.

1 Introduction

Artificial markets based on multi-agent simulation, as [3], emerged due to the characterization of financial markets as complex fractal systems. This field tends to explain stock market dynamics based on the reproduction of universal stylized facts (fat tails on returns distribution, absence of returns autocorrelation, volatility clustering...) observed on mature markets, as found by [5]. However, the important work of Bekaert and Harvey on the characteristics of emerging markets, [1], stated that emerging markets are mainly characterized by important asymmetric information due to the influence of local irregular channels of information transmission. They explained the high predictability of returns by the high correlation to local information and the low informational efficiency. Nowadays, aside from empirical studies, we remark an absence of works tending to explain these emerging markets stylized facts based on artificial markets. Thus, the aim of this work is to use behavioral and informationally heterogeneous agents to model immature market specificities, based on asymmetric information and investors herd behavior. Then, our work will be inspired by Pastore et al. (2010) to introduce an investors network modeling the information spread through sentiments. We will improve the modeling in [8] by introducing a scale-free directed weighted network based on [2] where agents represented by nodes

J. Bajo et al. (eds.), *Trends in Prac. Appl. of Agents, Multi-Agent Sys. and Sustainability*,
Advances in Intelligent Systems and Computing 372, DOI: 10.1007/978-3-319-19629-9_1

have mixed behaviors (fundamentalist, chartist and noise trading as in [4]). This will reproduce assortative[1] behaviors which will be linked, as explained in section 3, to asymmetric information.

This paper is structured as follows: section 2 presents the modeling of the considered traders behaviors that interact on a social network according to their beliefs. In section 3, simulations are carried out under our choice of network topology dynamics and agents' behaviors structure to determine the way of emergence of immature markets stylized facts. In section 4, concluding remarks are presented.

2 Behavioral Network for Simulating Markets with Asymmetric Information

We introduce a Behavioral and Informational Agents-Based Model (denoted BI-ABM) that models N heterogeneous agents who trade a single risky asset, having an initial price, $P_F = 10$, on a limit order-driven market. Agent i takes its investment decision by combining its trading behavior and the information shared with its neighbors on a behavioral network, through social sentiment.

2.1 Agent Characteristics

Each agent, i, is endowed by an initial cash, $C_{i,0} \to$ Unif.[0,50.P_F], an initial amount of owned assets $A_{i,0} \to$ Unif.[0,50] and an initial sentiment $S_{i,0}$. As presented by [4], agent i forms its trading behavior according to a mix of random weights $g_i^1 > 0$, $g_i^2 > 0$ and $n_i > 0$ representing respectively its degrees of fundamentalism, chartism and noisy trading. These random weights are generated, at the initialization, according to exponential probability distributions of respective means σ_1, σ_2 and σ_n (σ_n is fixed here to 0.1 modeling very low presence of noise traders). Finally, based on its behavior, agent i has a relative risk aversion, $\alpha_{i,t}$, that will evolve during time, and a time horizon τ_i reflecting its visibility, that is kept constant during time, measured as defined by [4]:

$$\alpha_{i,0} = \alpha^0(1+g_i^1)/(1+g_i^2) \text{ and } \tau_i = \tau^0(1+g_i^1)/(1+g_i^2) \text{ with } (\alpha^0, \tau^0) = (0.1, 100) \quad . \tag{1}$$

Based on Eq. (1), agents with higher fundamentalist component have more risk aversion and longer time horizon, giving them an additional ability to analyze market fluctuations and consequently better investment strategies than chartist agents. These properties make fundamentalists more informed than chartists.

[1] In an assortative network, attachments are established between nodes of similar specificities [6, 7].

2.2 Behavioral Network with Preferential Attachment

Stock markets are places where agents trade and share information about assets fluctuations, based on their social relationships, i.e. each agent exerts and undergoes social influence, through its perception and its strategic behavior. In this way, we assume as [8] that information flows can be represented by a directed weighted network, modeling the fact that social influences are asymmetric. However, the network modeling is improved here by considering a growth scale-free network able to reproduce assortative dynamics. We start from the growing directed scale-free network proposed by Bollobás et al. (2003) and introduce several improvements. Agent exerts influence through its sentiment and undergoes influence through social sentiment. For this reason the weight of each direct edge, from agent i to agent j, depends on the behavior of the influential agent, i.e. more the agent has a fundamentalist behavior more its influence on the financial market is important. Then, this relation weight is given by: $w_{ij} = g_i^1/(g_i^1 + g_i^2 + n_i)$. Second, based on the complex structure and connectivity on a financial market we assume that an agent may have multiple relations, which are reflected by an initial in-out edges number chosen randomly in $\{1, \ldots, m\}$. Thus, by increasing m, we will have more social influence and less rational investment decision. Our last contribution concerns the introduction of assortative dynamics as proposed by [7]: the new introduced node i tends to connect to the neighbor l of its neighbor j, where l is chosen randomly according to the following probabilities: $P(l) = (s_l^{\alpha})^{\theta} / \sum_{n \in \Gamma_l} (s_n^{\alpha})^{\theta}$ where s_n^{α} is the total weight of α-edges and $\alpha \equiv in$ in the case of out-edge from i to l and $\alpha \equiv out$ in the case of in-edge from i to l. θ is the variable that explains clustering on the network, i.e. lower is θ more probable are the clusters formation and more pronounced are the attachment between similar behavior agents. Γ_l denotes the neighbors set of node l. Therefore, by varying the couple (m, θ) we will be able to reproduce different assortative dynamics on the network. More details on the role of (m, θ) are given in Tab. 1. Therefore, we have four assortative measures $r_{\alpha,\beta}$ where $(\alpha, \beta) \in$ (in,out)-edge (see [6]): for $r_{\alpha,\beta} > 0$, we are talking about assortative network, while, for $r_{\alpha,\beta} < 0$ we have a disassortative network. For eg. in the case of in-in assortativity, agents with high in-edges weights prefer an attachment with agents with high in-edges weights, viceversa for in-in disassortativity. We will detail later (section 3) how assortativity can be linked to asymmetric information microstructure.

2.3 Market Design

The mechanism of BI-ABM algorithm is described through the following pseudo code (Algorithm 1.). We assume that independently from its social position and its wealth, at each step any agent has the chance to trade on the market. Then, through BI-ABM we choose randomly an agent to submit its investment desire. To take its decision, agent forms its sentiment towards the market activity based on its last strategy performance measure $U_{i,t}$ which depends on its last profit, $\pi_{i,t}$, and on the performance feedback of its in-neighbors $U_{i,t}^s$. In other words, agent i forms a perception based on its belief and on the social atmosphere.

Therefore, we support the ability of agent i to observe the general performance of its in-neighbors by referring to its ability to be informed through social channels on asset performance before taking its investment decision. The market sentiment has an impact on agent i risk aversion which changes according to its new perception. Finally based on its sentiment, its risk aversion and its time visibility τ_i, agent i desires a quantity $Q^d_{i,t}$ and a limit price $p_{i,t}$, that depends also on the current price p_t. By comparing the desired quantity to its actual portfolio, agent i takes its investment decision through an order submission.

Algorithm 1. BI-ABM pseudo code

```
Initialize(N, t = 0, T, P_F, σ_1, σ_2, σ_n, m, θ, α^0, τ^0)
for i = 1 : N do
    Simulate(A_{i,0}, C_{i,0}, g_i^1, g_i^2, n_i)
    Calculate(α_{i,0}, τ_i)
end for
Initialize(Behavioral Network)
while t < T do
    i ← Random(1, N)
    U_{i,t} ← Function(U_{i,t-1}, π_{i,t})
    S_{i,t} ← Function(U_{i,t}, U^s_{i,t})
    α_{i,t} ← Function(α_{i,t-1}, S_{i,t})
    p_{i,t} ← Function(p_t, S_{i,t}, τ_i)
    Q^d_{i,t} ← Function(C_{i,t}, p_t, α_{i,t}, τ_i)
    if Q^d_{i,t} > A_{i,t} then
        BuyOrder ← (Q^L_{i,t} = Q^d_{i,t} − A_{i,t}, p_{i,t})
    else if Q^d_{i,t} < A_{i,t} then
        SellOrder ← (Q^L_{i,t} = A_{i,t} − Q^d_{i,t}, p_{i,t})
    end if
    t ← Simulate(Time)
end while
```

2.4 Information Spread and Performance Measure

To improve the information modeling introduced by [8], we assume that information flows on the market due to the strategy performance measure defined by [3]. This measure helps agent i at time t to form its own market perception through a sentiment $S_{i,t} \in [-1, 1]$ quantified as follows:

$$S_{i,t} = \max\left(\min\left(a \cdot U_{i,t} + b \cdot U^s_{i,t}, 1\right), -1\right) \quad . \tag{2}$$

Following [8], a denotes the self confidence generated according to a uniform distribution Unif.[0,0.2] and b is the social confidence defined as $b = 0.6 - a$, where the sign of b is randomly changed according to probability 0.5 expressing that at different step agent may change its confidence towards the social perception (for $b > 0$, agent has a social confidence). Based on [3], the performance measure is given by: $U_{i,t} = \eta U_{i,t-1} + \pi_{i,t}$ where $\pi_{i,t}$ is the surplus of current transaction and η is the fundamentalist weight of agent i (higher is the fundamentalist weight, higher is the confidence about its past performance). Let $U^s_{i,t}$ denote the social influence exerted on agent i from its in-neighbors at time t. $U^s_{i,t}$ is measured by the weighted performances of agent i in-neighbors.

2.5 Simulating Order Decision

Agent i submits a limit order by defining its limit price, its investment quantity and its order nature.

Limit Order Price. Agent i begins by analyzing the market volatility, through the return variance $V_{i,t}$ along its time horizon τ_i. Taking into account its psychological variable $S_{i,t}$, the limit order price is given by: $p_{i,t} = p_t[1 + S_{i,t}].N(1, V_{i,t})$, where $N(.,.)$ is a simulation from a normal distribution.

Desired Quantity. Risk aversion is defined economically as a behavioral and psychological variable. We support that risk aversion can not be constant during time and will depend on the investor subjective perception on stock market. This is why it should be updated based on agent sentiment on BI-ABM, as follows: $\alpha_{i,t} = \alpha_{i,t-1} \times S_{i,t}$ if $S_{i,t} > 0$ and $\alpha_{i,t} = \frac{\alpha_{i,t-1}}{|S_{i,t}|}$ if $S_{i,t} < 0$. Then, we assume that agent i would invest its cash subject to its risk aversion and market volatility (risk measure). Higher is the market risk, lower is the desired quantity. Therefore, in BI-ABM we choose a simple form of $Q^d_{i,t}$ as: $Q^d_{i,t} = C_{i,t}/(p_t \times \alpha_{i,t} \times V_{i,t})$.

Limit Order Quantity and Order Nature. The limit order quantity $Q^L_{i,t}$ is defined as follows. Let $\Delta_{i,t} = Q^d_{i,t} - A_{i,t}$, if $\Delta_{i,t} > 0$, agent i submits a buy order $(Q^L_{i,t} = \Delta_{i,t}, p_{i,t})$. Conversely, if $\Delta_{i,t} < 0$, agent i submits a sell order $(Q^L_{i,t} = |\Delta_{i,t}|, p_{i,t})$. When executed, the current price is updated by $p_t = (a_t + b_t)/2$.

3 Simulation and Results

In this section different configurations of BI-ABM are simulated in order to reproduce earlier findings on chartists impact and asymmetric information effects and to conclude about the role of these two phenomena on reproducing immature markets stylized facts. The results will be the outcome of 100 000 seconds market simulation repeated 20 times with different random seeds. Then, we checked the model robustness by repeating the simulation with small variations of parameters values, without changing of agent behavior and network topology. The same general features exposed below were obtained supporting the robustness of the presented results.

3.1 Model Parameters and Reproduction of Mature Markets Stylized Facts with BI-ABM

To reproduce universal stylized facts, we calibrate our model by assigning all variables. Referring to the literature and to simulation experiments we found the artificial market configuration that reproduces stylized facts on price series and scale-free characteristics on simulated network. The Main BI-ABM parameters values are shown in Tab. 1. For $\sigma_2 = 2$ and $(m, \theta) = (3, 1)$, we generate universal stylized facts that validate our artificial stock market. The simulated returns distribution presents fat tails: based on Hill tail index[2], the negative tail index is equal to 3.73, while the positive tail index is equal to 3.95. This distribution has a negative skewness coefficient (-0.22). All these results are very close to the characteristics of real markets ([5] found that tail indices of real returns data are comprised generally between 2 and 5). Other stylized facts concerning the absence of return autocorrelation and the presence of long range memory on volatility were proved based on the respective values of Hurst exponent (based on the modified R/S statistics described in [4]), $\mathrm{H_r} = 0.48 \approx 0.5$ and $\mathrm{H_{|r|}} = 0.71 > 0.5$ (0.5 is the H value related to the absence of autocorrelation).

Table 1. Main BI-ABM Parameters Values

Parameters	Values	Comments
N	1000	Agents number on BI-ABM
σ_1	10	Fundamentalist presence
σ_2	$\{1, 2, 3\}$	Chartist presence: more chartist reflects more herd behavior
m	$\{1, 3, 10, 20\}$	Social connectivity: higher connectivity reflects more social influence and less rational decision
θ	$\{-3, -1, 1, 3\}$	Social clusters: lower θ stimulates connection between similar agents and reduces sharing information between heterogeneous agents

3.2 Reproduction of Immature Markets Stylized Facts with BI-ABM

Based on our tuning parameters σ_2 and (m, θ), we aim to find emerging markets stylized facts and explain their origins based on information asymmetry and herd behavior of chartist agents.

Emergence of Information Asymmetry Based on Network Topology. Our aim is to study the different network topologies by varying the parameters (m, θ). The simulated network topologies are then analyzed based on the assortative coefficients as explained in Subsect. 2.2. By analyzing assortative measures, we obtain the following results: r_{in-in}, $r_{out-out}$ and r_{out-in} reach

[2] Hill tail index measures the tail exponent: the lower it is, the fatter the tail of the distribution. In real markets the tail index is generally below 4.

their maximum positive values for $m = 20$ and $\theta = -3$ (assortativity) and their minimum negative values for $m = 1$ and $\theta = 3$ (disassortativity). Then, for the first and second measures, the increase of m and the decrease of θ stimulate chartists communities and fundamentalists communities. While, the third measure, r_{out-in}, shows more herd behavior due to the connection between more influential agents, fundamentalists, and more influenced agents, chartists. However, the analysis for r_{in-out} ($r_{in-out}^{m=1,\theta=3} = -0.023$ and $r_{in-out}^{m=20,\theta=-3} = -0.19$) reflects that more influenced agents (chartists) tend to connect to less influential agents (chartists), when m increases and θ decreases. This yields the emergence of communities of chartists. Thus, we prove that network topology with higher m and lower θ stimulates communities between similar behavior agents. Therefore, fundamentalists will upgrade their strategic decisions based on sentiments of their fundamentalists neighbors, while chartists will disrupt their strategic decisions due to the influences exerted by other low informed agents. Thus, we summarize the previous findings by linking the assortative typology to information asymmetry on BI-ABM, one of the most important characteristic of immature financial markets.

Emergence of Immature Markets Stylized Facts. Turning to literature, we consider four relevant immature markets stylized facts: heavier tails than those of mature markets and positive skewness on returns distribution, over predictability of return and longer memory on volatility. From the results analysis, we confirm, firstly, the results given by Subsect. 3.1. stipulating the reproduction of developed market stylized facts. Then, by moving parameters to a microstructure topology with more asymmetrical information and more herd behavior, we fall on immature markets stylized facts. Indeed, with high asymmetric information, $(m, \theta) = (20, -3)$, and important chartist behavior $\sigma_2 = 3$, we find a return distribution specified by a positive skewness of 0.15 and heavier tails, comparing to the mature market microstructure, with Hill indices of 2.7 and 2.75 for the positive and negative tails respectively. Furthermore, based on Hurst exponent we find on this market over predictability of returns, $\mathrm{H_r} = 0.53 > 0.5$, and excess long memory on volatility, measured as autocorrelation of absolute returns, $\mathrm{H}_{|\mathrm{r}|} = 0.84 > 0.5$. Therefore, based on BI-ABM we can move from mature market to immature market microstructure as summarized in Tab. 2. These results confirm that emerging markets stylized facts can be explained by asymmetric information and investor herd behavior.

4 Main Results and Conclusion

The principal goal of the developed BI-ABM was to improve the agents-based modeling of financial stock markets to reproduce and explain immature markets stylized facts. It is assumed that these stylized facts are caused mainly by asymmetric information and agents herd behaviors. Our main contribution was to exploit network topology for modeling information asymmetry and investors sentiments propagation, based on assortativity generated by agents behavior.

Table 2. BI-ABM from Mature Market Microstructure to Immature Market Microstructure

Parameters	Microstructure	Negative Tail	Positive Tail	Skewness	H_r	$H_{\|r\|}$
$(m, \theta) = (3, 1); \sigma_2 = 1$	Symmetric typology and low chartist.	3.80	3.93	−0.12	0.46	0.76
$(m, \theta) = (10, -1); \sigma_2 = 2$	Asymmetric typology and moderate chartist.	3.43	3.33	0.09	0.49	0.79
$(m, \theta) = (20, -3); \sigma_2 = 3$	High asymmetric typology and high chartist.	2.75	2.70	0.15	0.53	0.84

These modeling contributions allowed us to prove that asymmetric information and herd behavior can reproduce, in agents-based artificial markets, features characterizing immature financial markets.

References

[1] Bekaert, G., Harvey, C.: Emerging Markets Finance. Journal of Empirical Finance 10, 3–55 (2003)
[2] Bollobás, B., Borgs, C., Chayes, T., Riordan, O.: Directed Scale-Free Graphs. In: Proc. 14th ACM-SIAM Symposim on Discrete Algorithms, pp. 132–139 (2003)
[3] Brock, W.A., Hommes, C.H.: Heterogeneous Beliefs and Routes to Chaos in a Simple Asset Pricing Model. Journal of Economic Dynamics and Control 22, 1235–1274 (1998)
[4] Chiarella, C., Iori, G., Perello, J.: The Impact of Heterogeneous Trading Rules on the Limit Order Book and Order Flows. Journal of Economic Dynamics and Control 33, 527–537 (2009)
[5] Cont, R.: Empirical Properties of Asset Returns: Stylized Facts and Statistical Issues. Quantitative Finance 1, 223–236 (2001)
[6] Foster, G.J., Foster, D.V., Grassberger, P., Paczuski, M.: Edge direction and the structure of networks. PNAS 107(24), 1–6 (2010)
[7] Guo, Q., Zhou, T., Liu, J.-G., Bain, W.-J., Wang, B.-H., Zhao, M.: Growing Scale-Free Small-World Networks with Tunable Assortative Coefficient. Physica A 371, 814–822 (2006)
[8] Pastore, S., Ponta, L., Cincotti, S.: Heterogeneous Information-Based Artificial Stock Market. New Journal of Physics 12, 053035 (2010)

In Whose Best Interest? An Agent-Based Model of High Frequency Trading

Paolo Pellizzari

Dept. of Economics, Ca' Foscari University, Venice
paolop@unive.it

Abstract. We study a stylized model of High Frequency Trading in which traders equipped with private values and costs operate in a Continuous Double Auction. They can revise their orders with different frequencies and, hence, (only) some agents can repeatedly revise and resubmit orders in the same session, mimicking the behavior of high frequency traders. All agents attempt to maximize profits, learning which bid and ask is to be posted in a given configuration of the book. We analyze the efficiency of the resulting market and the way the surplus from trading is apportioned among agents as a function of the number and type of high frequency traders. We find that the presence of a small proportion of high frequency traders increases the overall efficiency of the market; secondly, the ones who have the chance to frequently revise the offers learn to extract a disproportionate fraction of the profits that ordinarily would belong to slow traders.

1 Introduction

High Frequency Trading (HFT) broadly refers to trading strategies involving fast submission, cancelation and revision of orders in a Continuous Double Auction (CDA). Often, a massive number of orders is submitted to be left on the book for very short times and it is estimated that more than 50% of daily stocks are exchanged by HFT financial firms. There is anecdotal evidence that such firms are willing to spend hundreds of millions to increase the speed of their operations by a few milliseconds through the installation of fiber cables in the ocean or purchasing other related IT infrastructure, (Philips, 2012).

The remarkable surge of HFT in recent years was investigated to assess whether and to what extent such practices can provide liquidity to slow and more traditional financial operators, alter the volatility of the market, (Biais et al., 2014) or contribute to the destabilization of the regular functioning of exchanges in some specific instances, (Easley et al., 2011).

Clearly, there are issues worth investigating for the policy maker: HFT may improve the liquidity in the market, as the order book is flooded with thousands of bids or asks. In particular, while market orders consume liquidity, plenty of limit orders are typically interpreted as a way to provide liquidity or a way to manage trading akin to what an ordinary market-maker would do. However, it was insightfully pointed out by (Hasbrouck and Saar, 2009,

J. Bajo et al. (eds.), *Trends in Prac. Appl. of Agents, Multi-Agent Sys. and Sustainability*, Advances in Intelligent Systems and Computing 372, DOI: 10.1007/978-3-319-19629-9_2

Hasbrouck and Saar, 2013) that the picture is more blurred: a limit order certainly offers an option to trade to others but, if this opportunity is (only) available for a fraction of a second, it may be difficult for the "ordinary" trader to use it and, indeed, the triggered order may reach the market too late or generate other unexpected and potentially negative effects. Moreover, HFT may generate intermittent volume and suddenly reduce the number or orders that are posted, thus enlarging the spread and failing to be reliable source of liquidity.

In this paper we describe a stylized model of HFT in which a set of agents want to buy or sell a single stock in a trading session. This can be done submitting limit orders, with no certain execution, or marketable orders that instead immediately hit the opposite side of the book and are executed at less advantageous prices. Every trader maximizes his expected profits and must decide the limit price for a single unit of the stocks. We assume that the limit price is a simple linear combination of the outstanding best aks and bid at the time of submission, as in (Ladley and Pellizzari, 2014).

HFT is included in the model allowing some traders to cancel and resubmit multiple times during a trading session: in contrast to the ordinary agent, who can send an order only once, the fast trader will have f additional chances to cancel the previous order (if it was lying in the book unexecuted) and revise the submission posting a fresh quote or a market order. Clearly, this permits to exploit orders that appeared after the first submission or, say, to change the degree of aggressiveness of the order with the conditions of the market. The value of f spans the speed of the traders and while ordinary, slow traders have $f = 0$ (no further chance beyond the unique one granted to everyone in the session), larger values of f are meant to model various intensities of the *high frequency* attribute.

We are mainly interested in quantifying the efficiency, if any, that the HF traders would bring with respect to the same market with slow agents. As a second target we aim at describing the way profits from trading are distributed among traders, when a fraction of the agents are engaged in HFT and with a special concern for the (differential) effects on the remaining slow traders.

In the next Section, we present the model describing the agents, the order book and how strategies are learned. Section 3 uses simulations to show results related to the efficiency of the market with/without HFT and how profits from trading are shared. Finally, we close with some additional remarks.

2 The Model

2.1 The Traders

We consider a standard Continuous Double Auction (CDA) where at each time step a single trader enters the market. An equal number N of buyers and sellers are endowed with a positive reservation value $v_i, i = 1, ..., N$ or positive cost $c_i, i = N+1, ..., 2N$. In every trading session they can buy or sell a single unit of stock and, if a transaction occurs at a price p, the profit π_i awarded to the agent is $v_i - p$ or $p - c_i$, respectively. As no profit is made in the lack of a trade, i.e.,

$\pi = 0$, traders have the obvious incentive to exchange their unit, bidding (asking) less (more) than the value (cost) they are endowed with. Scholars have discussed other mechanisms to model trading: in (Rosu, 2009), traders are equipped with a discount rate that penalizes delayed or missed trading differently, allowing to define sets of *patient* or *impatient* traders. We prefer to use private values and costs because an equilibrium price is easily computed and still impatience is immediately and intuitively related to the magnitude of values and costs. In fact, buyers with high values or sellers with low costs suffer large losses if they do not trade and can be thought as impatient traders; conversely, buyers/sellers with values and costs close to the equilibrium price can generally obtain small gains and are nearly indifferent between trading and non-trading, hence resembling *patient* traders.

We deal with cancellation in a simplified and non-strategic way: at the end of every time step each order stored in the book is cancelled[1] with (a small) exogenous probability $P_c > 0$ that is independent of time, state of the book and of the specific agent acting in that period. This is the only source of cancellation for traders whose $f_i = 0$. Fast traders with $f_i > 0$ can have their order exogenously canceled as just described or intentionally cancel (and resubmit) their order up to f times after the first submission, if they have not traded yet.

2.2 The Book

At any time t the book is a double sequence of outstanding unit orders

$$S_t = \{0 \leq ... \leq b_{3t} \leq b_{2t} \leq b_{1t} < a_{1t} \leq a_{2t} \leq a_{3t} \leq ...\},$$

where $b_{1t}, b_{2t}, ...$ and $a_{1t}, a_{2t}, ...$ are the lists of buy and sell orders in the books. We often omit the time index for simplicity. The highest bid b_1 and lowest ask a_1 are referred as best bid and best ask, respectively.

Traders submit a single order when they enter the market in a random period, according to the rules mentioned in the previous subsection. As the quantity is fixed at one unit, the trader must decide the limit price that is computed using a function of the state of the book and his valuation/cost.

The submission of an (unit) order with limit price l changes the book and results in an immediate trade, a *marketable order*, if the bid (ask) is greater (smaller) than or equal to the best ask (bid). In this case, the two agents involved in the transaction get the associated profits and the book is then updated, changing one between the best bid or ask. If instead the new order is not marketable, it is inserted in the book, maintaining its ordering, to be possibly used in future trades. Any profit occurring after the time of submission t is accrued in the same way to the parties involved whenever a transaction takes place.

2.3 The Strategy

In real markets, traders can decide the time at which to submit and the limit price l they want to use (in our model the quantity is fixed at one). The time of first

[1] We never cancel the order in the time step in which it is submitted.

submission is random and drawn independently from any other variable of the model. We further assume, as in (Ladley and Pellizzari, 2014, Pellizzari, 2011), that each agent develops his own trading strategy picking the limit price using the relation

$$l_{it} = \alpha_i a_{1t} + \beta_i b_{1t} + \gamma_i, \tag{1}$$

where α_i, β_i and γ_i are individual constant to be found maximizing expected profits and a_{1t}, b_{1t} are the best ask and bid at the time t in which the order[2] is issued. Agents for which $f_i > 0$ have multiple chances to cancel the orders: in more detail, in f random periods after the first submission, they can cancel their unexecuted order and resubmit at different limit prices (that are computed using (1) with the outstanding best bid and ask at the time of resubmission).

In other words, the model assumes that HFT really is only a matter of how often one can cancel the order, reevaluate his/her limit price and resubmit the order. We are aware that this simple approach has limitations and real HF traders may be more sophisticated in several ways. However, it is clear that having the chance to cancel and resubmit at high frequency should give an edge to some traders who can, say, wait-and-see for better market conditions or have a much higher chance to selectively pick the most beneficial quotes left on the book for some time. As pointed out by a referee, traders have no budget constraints and this is a limitation of the present treatment. We believe results are not too sensitive to this aspect, provided that all traders can buy/sell the same quantity as done in this model. Notice that even agents who can revise their quotes often can still buy/sell only once per trading session.

The constants α_i, β_i and γ_i in (1) are individually tuned using a Evolution Strategies (ES) numerical scheme to maximize expected profit (hence, many sessions are averaged and as a consequence we also smooth the effect of the random position in the submission queue). The ES method belongs to the family of evolutionary algorithms and superficially resembles the well known Genetic Algorithms but requires no binary-coding or discretization of $\alpha_i, \beta_i, \gamma_i$ and has self-adapting meta-parameters that can be used to gauge whether convergence has been reached, see (Beyer and Schwefel, 2002) for an overview.

With no hope to be exhaustive due to lack of space, maximization of profits is jointly performed by all the traders who learn in pools with the same value/cost and f: they periodically rank their strategies and deterministically discard the least performing half. Within each pool, the survived strategies are crossed, mutated and tested in the market searching for additional improvements. ES determine a numerical equilibrium in which no agent has further incentive to alter his/her constants. In such a sense, the outcome is (just) one of the multiple trading equilibria that can appear given the values of v_i, c_i, f_i. We do not investigate here the number and properties of different equilibria and are content with the observation that, whatever they are, the selected triples $\alpha_i, \beta_i, \gamma_i$ form a stable configuration that is *learnable* as proved by the fact that is was indeed

[2] We do not model an initial fixing mechanism and assume the best 10 bids/asks of the previous session are available in the "opening" phase of the next one. We never observe occurrence of an empty book in any of our simulations.

built by the ES optimizing procedure, that can be thought as a learning device in which traders (and their strategies) compete as in the marketplace.

3 Results

3.1 Implementation and Parameters

We simulated the model described in the previous section using $N = 120$, effectively allowing 240 traders to participate in every trading session. To assess the variability inherent in any simulation, we have run 20 independent simulations in which 5000 trading sessions take place. Expected profits are computed averaging 100 trading sessions, after which ES performs selection, recombination and the so-called s-mutation (for meta-parameters) and y-mutation (for trading strategies).

All the results are based on the average of 20 simulations, in which we keep the last trading session (out of the 5000 that were used to "train" the traders towards optimal extraction of profits). Values and costs are symmetric around 10, which is therefore the theoretical equilibrium price and cancelation probability P_c is set at 0.01.

We performed 5 experiments to investigate how HFT affects efficiency and the way profits are apportioned among traders. Table 1 describes our design in which, starting from a benchmark case of slow traders alone ($f = 0$), we progressively add sets of HF traders ($f = 3$) with different values and costs. In all the experiments there are 240 agents, the *ex ante* equilibrium price is 10 (this is unknown to traders) and we vary the type and the fraction of HF agents. It is convenient in the following to label agents according to their values and costs: in particular, we nickname marginal agents whose value or cost is 10 as *calm*, referring to the fact that they have nothing to loose if they fail to trade at the equilibrium price; using the same intuition, buyers or sellers with $v = 14$ or $c = 6$ are *nervous*, as they feel more the need to trade; finally, when values reach 18 and costs as are low as 2, traders are named as *hysteric*, to capture the fear for sizable losses in case of no trade. We remark that the terms *calm, nervous* and *hysteric* are related to personal values and costs and have nothing to do with the high frequency feature that depends on f. Experiments 2, 3 and 4 increases the number of HF traders of a single type, respectively, still maintaining the fraction of HFT at 40/240. Experiment 5 considers a situation where 50% of the population adopts HFT.

The last column of Table 1 shows the median allocative efficiency reached in the experiments. Observe that when no HFT is present, agents nevertheless learn to extract more than 95% of the trading surplus (median value over 20 simulations). As more HF traders are added, efficiency increases reaching a median value of 100% in Experiment 3, in which half of the hysteric traders use HFT. The effect is smaller if calm or nervous HFT is included in the market and the same can be said for Experiment 4 when HF traders belong to all types (calm, nervous and hysteric). We guess that this retreat in efficiency when many

Table 1. Description of the 5 experiments. The first row describes the benchmark case with no HFT and each listed value/cost represents a set of 20 traders. Groups of HF traders are boldfaced and starred and, for instance, in Experiment 3 there are 20 *hysteric* HF buyers (sellers) with value (cost) equal to 18 (2). The last column shows the efficiency of the market (median value of 20 runs).

	HFT	Buyers' values v_i						Sellers' costs c_i						Eff
Benchmark	0	18	14	10	18	14	10	2	6	10	2	6	10	0.955
Experim. 1	40	18	14	10	18	14	**10***	2	6	10	2	6	**10***	0.963
Experim. 2	40	18	14	10	18	**14***	10	2	6	10	2	**6***	10	0.986
Experim. 3	40	18	14	10	**18***	14	10	2	6	10	**2***	6	10	1.000
Experim. 4	120	18	14	10	**18***	**14***	**10***	2	6	10	**2***	**6***	**10***	0.963

HF traders are present may be due to the additional complication (and related errors) of learning the optimal trading strategy in such a hectic environment.

Efficiency is a global measure of the ability of traders to transact when it is convenient given the individual valuations and the equilibrium price driven by the demand and supply for the asset. However, it says nothing about the precise way in which the available surplus is shared among traders. Table 2 shows how profits are distributed in our experiments. The profits of the first row, relative to the benchmark, are used to normalize the entries of the other rows and cases where HF traders are in action are boldfaced.

In Experiment 1, we see the effect the introduction of calm HF traders. With respect to the benchmark case, the average profit of all calm traders increases by almost 22%, while calm and hysteric traders virtually keep the same share as in the benchmark case.

When HFT is unleashed for some nervous traders (Experiment 2), they see their own profit increase by 13%, slashing by about 43% the share of the calm agents.

Table 2. Fractions of profit gained by calm, nervous and hysteric traders when HFT enters the market. All figures are normalized with respect to the benchmark case with no HFT.

	Calm	Nervous	Hysteric
Benchmark	1.000	1.000	1.000
Experim. 1	**1.218**	0.993	0.999
Experim. 2	0.574	**1.129**	1.005
Experim. 3	0.700	0.999	**1.089**
Experim. 4	**0.171**	**1.078**	**1.028**

The profit of all hysteric traders, slow and fast, increases overall by nearly 9% in Experiment 3, at the expenses of a reduction of 30% for the calm trader with nervous traders again unaffected.

At least two things are worth noticing: firstly, adoption of HFT in isolation increases the profits of the agents that use this trading approach. Effects are relatively large for calm traders (+22%), even though the absolute magnitude of the variation is small (being traders' valuations equal to the equilibrium price), and are smaller but sizable in absolute terms for hysteric traders. Secondly and quite remarkably, if HFT is used by *one* subgroup alone, negative effects on profits are detectable mainly for calm (i.e., marginal) traders

In a nutshell, *HFT benefits the ones who use it but can damage some of the other agents that are worse off*, despite the increment of the efficiency that was shown in the last column of Table 1. If the model realistically depicts the market, traders may increasingly be pushed to employ HFT, producing the outcomes shown in "Experiment 4". In this case, in which agents equally split between slow and HFT, the profits of the calm traders are nearly wiped out (-83%), whereas both nervous and hysteric traders keep additional gains.

4 Discussion

In this paper, we present a simple model of HFT where some traders have in every trading session $f > 0$ additional chances to cancel and resubmit yet unexecuted orders. The higher f, the more frequently traders can adjust their orders, if needed, and exploit transitory opportunities in the book. In contrast, traditional (i.e., slow) traders have a single chance to deposit an order per trading session, with no possibility to alter it.

Despite the simplicity of the model, we obtain two main insights: the introduction of HFT marginally increases the market efficiency and, hence, the surplus extracted is larger than in the presence of slow traders alone. In this sense, the pie gets slightly bigger as efficiency steps up from 95 in the benchmark case to 98% or more when few HF traders are introduced. We also find that HFT benefits the investors who are using it at the expenses of other traditional investors that have their profits eroded significantly or slashed.

In (Bernales, 2014) it is claimed that algorithmic traders with a trading speed advantage only reduce global welfare and this statement is to some extent at odd with our first finding. We instead corroborate, in our agent-based setup, Bernales' result that HFT *de facto* predates some of non-HFT agents, seizing a portion of their theoretical gains. One novel insight of our model follows the observation that, under the dynamics that we have just described, many traders would be tempted to invest in IT infrastructure and actually jump in the HFT arena. As a consequence, efficiency would increase less than when there is limited HFT. Hence, when too many HF traders operate in the market, for some the profit would shrink again. In a situation resembling a "prisoner dilemma", policymakers may attempt to regulate access to and activities of HFT firms, to curb arms races potentially able to waste costly private investments and hamper public welfare.

Finally, more realistic models may add to the this picture the explicit consideration that HFT is likely to be capable of processing information more accurately and faster than the ordinary trader. Such a research avenue could prove whether there are beneficial effects due to HFT in terms of, say, information diffusion and price discovery.

References

Bernales, A.: Algorithmic and high frequency trading in dynamic limit order markets. Tech. rep., SSRN (2014)

Beyer, H.G., Schwefel, H.P.: Evolution strategies: a comprehensive introduction. Natural Computing 1, 3–52 (2002)

Biais, B., Foucault, T., Moinas, S.: Equilibrium fast trading. Tech. rep., SSRN (2014)

Easley, D., Lopez de Prado, M., O'Hara, M.: The microstructure of the flash crash: Flow toxicity, liquidity crashes and the probability of informed trading. The Journal of Portfolio Management 37(2), 118–128 (2011)

Hasbrouck, J., Saar, G.: Technology and liquidity provision: The blurring of traditional definitions. Journal of Financial Markets 12(2), 143–172 (2009)

Hasbrouck, J., Saar, G.: Low-latency trading. Journal of Financial Markets 16(4), 646–679 (2013)

Ladley, D., Pellizzari, P.: The simplicity of optimal trading in order book markets. In: Dieci, R., He, X.Z., Hommes, C. (eds.) Nonlinear Economic Dynamics and Financial Modelling, pp. 183–199. Springer International Publishing (2014)

Pellizzari, P.: Optimal trading in a limit order book using linear strategies. Working Papers 16, Department of Economics, University of Venice "Ca' Foscari" (2011)

Philips, M.: High-speed trading: My laser is faster than your laser. BloomberBusinessweek (2012)

Rosu, I.: A dynamic model of the limit order book. Review of Financial Studies 22(11), 4601–4641 (2009)

Agent-Based Model for Order Routing and Financial Market Integration

Andrew Todd, Peter Beling, and William Scherer

University of Virginia, Charllotesville, VA, USA
{aet6cd,pb3a,wts}@virginia.edu

Abstract. Agent-based modeling is alternative framework for providing explanations of the behavior of financial markets. In this paper, we consider a simple agent-based model designed to understand the effects of smart order routing in a multi-market setting. The goal is to understand how the prevalence of smart order routing results in different levels of market integration. We find that only a small percentage of smart order routing results in a remarkable level of market integration.

1 Introduction

The law of one price and arbitrage-free markets are fundamental notions in financial economics. The law of one price dictates that securities with identical cash flows will trade at identical prices. Significant deviations should be swiftly resolved by the strategic behavior of arbitrage traders [1].

From an empirical standpoint, we can observe cross-listed securities and develop measures of market integration based on mid-quote deviations and other proxies. However, we can only hypothesize about the forms of strategic behavior that leads to those outcomes. Agent-based models provide a framework in which to connect strategic behavior to market integration outcomes.

We introduce an agent-based model designed to study the effects of strategic order routing on market integration. The basic ingredients of the model include passive participants that place and cancel limit orders relative to the prevailing mid-quote, and liquidity-demanding participants, which we refer to as aggressive participants, that arrive randomly throughout the trading period. The market consists of two independent limit order books. Aggressive traders have the capability of routing their market orders to either book according to the prevailing quotes. The probability that such a trader "smart routes" an order is varied as part of the experimental design.

The power of agent-based models (i.e. their ability to capture specific details of market mechanisms and trading behavior) is also a source of problems. ABM allow many degrees of freedom [2]. In other words, for any model that one puts forward as an explanation of an empirical phenomenon, there may exists a number of other specifications in which the same behavior arrises. With this in mind, we employ a parsimonious specification of strategic behavior.

J. Bajo et al. (eds.), *Trends in Prac. Appl. of Agents, Multi-Agent Sys. and Sustainability*,
Advances in Intelligent Systems and Computing 372, DOI: 10.1007/978-3-319-19629-9_3

The structure of this paper is as follows. First, we provide a brief review of the pricing mechanism, the limit order book. We then provide a discussion of market fragmentation and automated trading. Finally, we present the model details and results of the computational experiment.

2 Market Mechanism

The pricing mechanism employed in our model is a limit order book. In some markets, designated intermediaries establish prices at which other market participants may trade. In this model, we employ an order-driven market, where the market participants themselves establish prices by submitting limit orders [3]. A limit order is an instruction to buy or sell a specified quantity of a given security at a specified price. In practice, limit orders are stored in two price-time priority queues known as the limit order book. Each limit order consists of a side, a quantity and a price. A bid with price $p > \hat{p}$ has priority over another bid with price $\hat{p}$. Similarly, an ask with price $p < \bar{p}$ has priority over another ask with price $\bar{p}$.

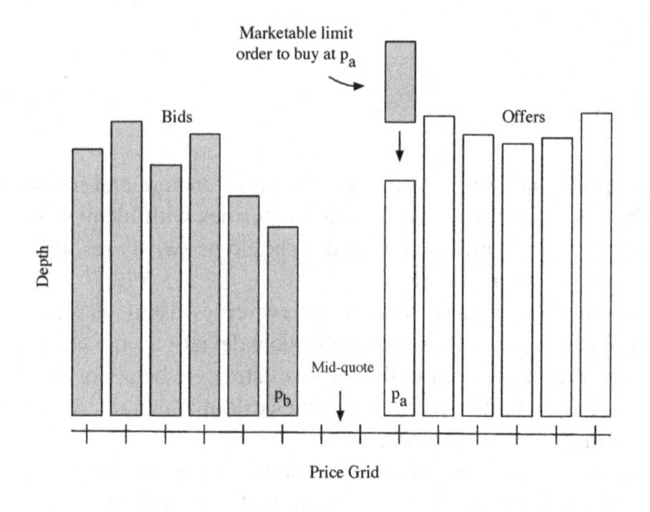

Fig. 1. Diagram of a limit order book

Best bid and ask prices are denoted p_b and p_a, respectively. In practice, market participants can submit market orders that execute against the best available limit orders. In our model, market orders are implemented as marketable limit orders, e.g. a new limit order to buy in which the price is greater than or equal to the best ask. In figure 1, we depict a limit order book on an integer price grid. A market order to buy would execute at p_a. A new limit order to sell at $(p_a - 1)$, would improve the best ask price. The tick size in the smallest price increment for orders.

Limit order markets may have a number of idiosyncratic features, including orders that have special behavior. In our model, we implement only the core features of a limit order book. The market mechanism, while often replaced with an abstraction for modeling purposes, is modeled explicitly in our agent-based framework. While the market mechanism itself is an important part of the market, we also need to account for the broader market structure, which we review briefly in the next section.

3 Market Fragmentation

Securities are often listed at multiple venues. The United States stock market is highly fragmented. Regulation NMS (National Market System), however, strictly dictates how quotes are integrated across markets. For example, Reg NMS protects orders against trade throughs. A trade through occurs when a transaction does not take place at the best possible price as currently quoted. Prior to the introduction of Reg NMS, more than 1% of volume traded through better quotes on U.S. equity markets [4]. U.S. securities, however, are also listed abroad taking the form of American Depository Receipts. An ADR trading in Europe or Asia is not an NMS security and is not given trade-through protection.

Europe has also opened the way for increased competition where securities are listed across markets both nationally and internationally [5]. For example, consider Nordea, a European bank. Nordea trades in Stockholm, Helsinki, and Copenhagen. Nordea shares trade at all three market centers in their respective currencies. And, at each location, Nordea is traded on a number of competing platforms. The market for Nordea is obviously fragmented, but it isn't segmented. The prices are connected by the strategic behavior of market participants.

Globally, the computerization and automation of exchanges has lead to a decrease in operating costs and the proliferation of a variety of alternative venues [6]. Market participants often face the decision as to where and to what extent they will trade at each venue. While some may choose to monitor and route orders to multiple venues, economic considerations may limit a number of participants to a single venue. Due to the time sensitive nature of such questions, market participants often relegate the decision to an automated trading system.

4 Automated Trading and Smart Order Routing

Market participants automate a wide variety of trading strategies. In this paper, we focus on automated trading strategies related to the execution of large orders, which we will refer to as parent orders. The parent order may be equivalent in size to a significant portion of daily trading volume. The trader tasked with executing the order is interested in minimizing costs including implicit costs such as the price impact of the trade. The trader may also have additional constraints, for example, a deadline (trading horizon).

The most common strategy is to split the parent order in to child orders and execute them throughout the trading horizon. To do so, the trader will assess the difficulty of the trade, based on its size and trading horizon, and select a strategy, which is generally associated with a particular benchmark. That strategy often involves routing the child orders to a number of different venues based on fees, spread, latency and other factors, which is often referred to as smart-order routing.

Smart order routing, in its most basic form, simply involves sending a child order to the most favorable venue. That venue may have better quotes, smaller fees, or other features that result in reduced transaction cost. Smart order routers also take into account the probability of a successful execution based on latency and other factors. For example, a limit order joining the best bid could result in an execution if the market moves against you between the time you submit and the time the order hits the matching engine. In the model presented below, we simply consider smart-order routing as routing an order to the limit order book with the most favorable quote (as seen in figure 2).

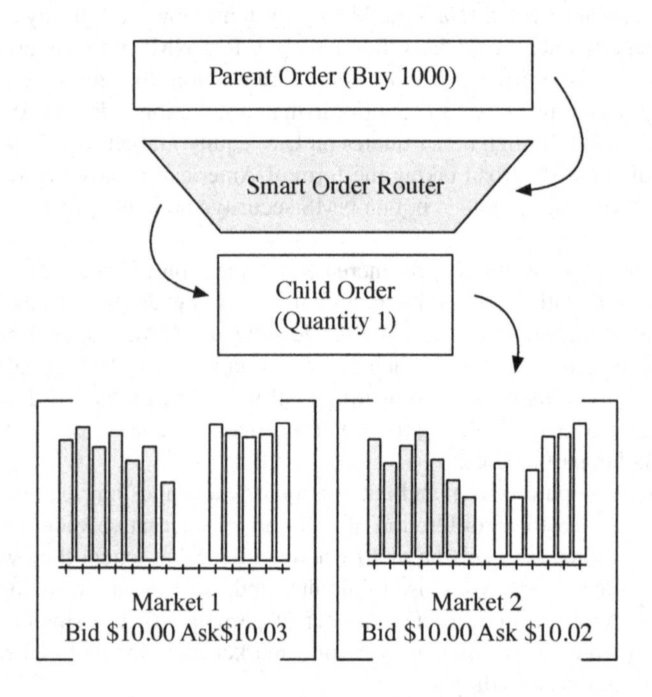

Fig. 2. Smart order routing as it is defined in the model. A parent order routes a child order to the market with the best offer.

5 Model

Our model is related to a line of work that treats the arrival of orders as a statistical process [7,8,9]. It is also related to models accounting for multiple trading venues [10,11,12]. The contribution of this paper is to connect the strategic behavior of order routing to levels of market integration. The model consists of two classes of participants: passive participants and aggressive participants. The model is discrete time. Agents are activated uniformly at random. Passive traders place limit orders uniformly at random relative to the opposite quote. The limit orders of passive traders are to buy or sell with equal probability. Bids take a price from the discrete set $\{p_a - \delta, \ldots, p_a\}$, and asks take a price from $\{p_b, \ldots, p_b + \delta\}$. The parameter δ governs the size of the price range. All orders have unit quantity. Orders that are not executed after n epochs are canceled.

Aggressive traders place market orders according to their current parent order, which consists of a quantity and a side (q, s). Each aggressive trader has an initial parent order. The quantity is drawn according to Zipf's Law [13]. The idea is that the power law distribution of the quantities of parent orders causes the long range dependence observed in the sides of market orders [14,15]. In our model, the side of the parent order is to buy or sell with equal probability. When an aggressive trader is activated it trades against its current parent order. A child order of size 1 is broken off and the parent quantity is updated to $q-1$. If the parent order is exhausted $(q=0)$ when an aggressive trader is activated, a new parent order is generated and the agent places a trade.

The orders of aggressive traders are smart-routed with probability λ and routed to the traders home market with probability $(1-\lambda)$. This parameter can be interpreted as the percentage of the parent order that is given to a broker with smart routing capabilities. Market participants often spread large orders over a number of competing brokers. Smart routing simply means sending the child order to the market with the most favorable quote, e.g. if the order is to buy the order is routed to market with the lowest offer. In the event of a tie, the order is routed to a market uniformly at random. Aggressive traders employ market orders and execute against the prevalent quotes at the time of each child order.

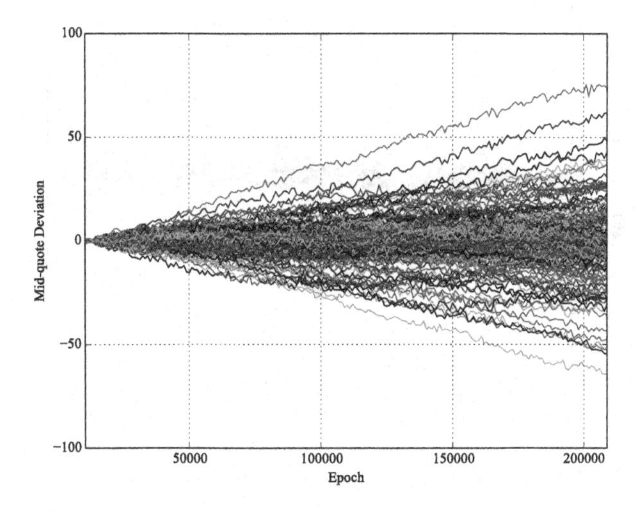

Fig. 3. Mid-quote deviations for 200 replications with $\lambda = 0.10$

6 Results

The computational experiment consists of two identical limit order markets, market 1 and market 2. Each market has 100 participants. There are 90 passive participants and 10 aggressive participants. Passive participants only trade on their home market. The aggressive participants have a home market, but have access to both. Each replication of

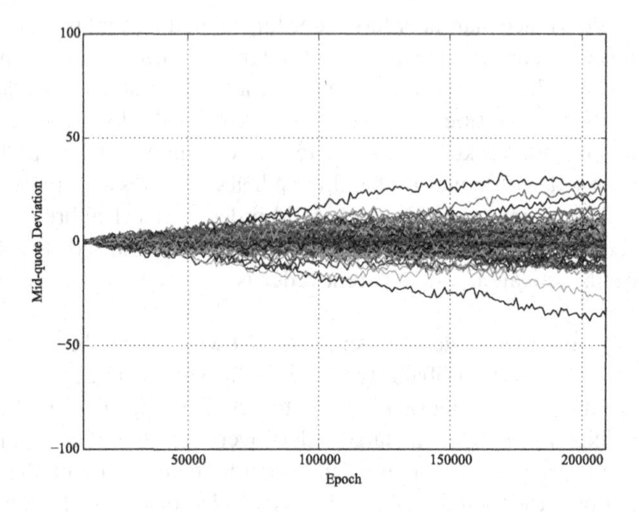

Fig. 4. Mid-quote deviations for 200 replications with $\lambda = 0.20$

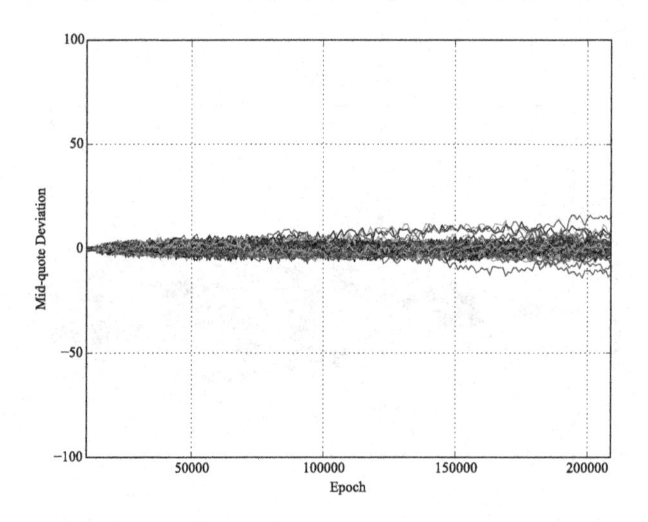

Fig. 5. Mid-quote deviations for 200 replications with $\lambda = 0.30$

the simulation lasts for 200,000 epochs not including a 10,000 epoch warm up period in which only passive agents trade. Limit orders are placed within 10 ticks of the prevailing opposite quote ($\delta = 10$) and limit orders that have not been executed for 10,000 epochs are canceled ($n = 10,000$). The probability of smart order routing is varied between 0.1 and 1.0 by increments of 0.1. The quantity of parent orders is drawn according to Zipf's Law with $\alpha = 1.75$.

Market integration is measured by examining the mid-quote deviation, where the mid-quote is defined as

$$\frac{(p_a + p_b)}{2} \tag{1}$$

The mid-quote deviation is $(m_1 - m_2)$ where m_1 is the mid-quote for market 1 and m_2 is the mid-quote for market 2.

Table 1. Minimum and maximum mid-quote deviation over 200 replications. Var. reports the sample variance of the mid-quote difference as measured in the last step of each replication.

λ	Min.	Max.	Var.
0.10	-65	77	323.75
0.20	-30	33	41.47
0.30	-14.5	17	6.40
0.40	-8	6.5	2.39
0.50	-6	5	1.45
0.60	-5	5	1.23
0.70	-5	3.985	1.07
0.80	-5	4.5	0.85
0.90	-4	5	0.92
1.00	-5	5	0.75

As can be seen in figures 3 - 5, moderately low levels of smart routing result in markets that are fairly integrated as measured by mid-quote deviations. Table 1 reports the minimum and maximum deviation over all 200 replications for each level of smart-order routing. While markets are fairly integrated for $\lambda > 0.40$, there are still periods of dislocation resulting from large order flow imbalances. The model provides a starting point for exploring more sophisticated order routing and additional types strategic behavior.

Ultimately, the level of market integration is a result of market structure, regulations, fees and other costs faced by traders. Realistic market structures involve asymmetric volume levels across venues, foreign exchange considerations, varying tick sizes and other complications. Agent-based models provide a framework for exploring dynamic aspects of more complicated market structures. We look forward to exploring additional features in future work.

References

1. Lamont, O., Thaler, R.: Anomalies: The Law of One Price in Financial Markets. J. Econ. Perspect. 17(4), 191–202 (2003)
2. LeBaron, B.: Agent-based Computational Finance. Handbook of Computational Economics 2, 1187–1233 (2006)
3. Puneet, H., Schwartz, R.: Limit Order Trading. Journal of Finance, 1835–1861 (1996)

4. Securities Exchange Commission: Regulation NMS, 43–45
5. Preece, R.: The Structure, Regulation and Transparency of European Equity Markets under MiFID. CFA Institute Position Paper (2011)
6. O'Hara, M., Ye, M.: Is market fragmentation harming market quality? J. of Fin. Economics, 459–474 (2011)
7. Smith, E., Farmer, J., Gillemot, L., Krishnamurthy, S.: Statistical theory of the continuous double auction. Quantitative Finance 3, 481–514 (2003)
8. Szabolcs, M., Farmer, J.: An empirical behavioral model of liquidity and volatility. J. Econ. Dyn. Control. 34, 200–234 (2008)
9. Mastromatteo, I., Toth, B., Bouchaud, J.: Agent-based models for latent liquidity and concave price impact. Physical Review E. 89, 042805 (2014)
10. Wah, E., Wellman, M.: Latency Arbitrage, Market Fragmentation, and Efficiency: A Two-Market Model. In: Proceedings of the 14th ACM Conference on Electronic Commerce, pp. 855–872 (2013)
11. Mo, S., Paddrik, M., Yang, S.: A study of dark pool trading using an agent-based model. In: Proceedings of the Conference on Computational Intelligence for Financial Engineering & Economics, pp. 19–26 (2013)
12. Huang, W., Chen, Z.: Modeling regional linkage of financial markets. J. Econ. Behav. Organ. 99, 18–31 (2014)
13. Newman, M.: Power laws, Pareto distributions and Zipf's law. Contemporary Physics 46(5), 323–351 (2005)
14. Lillo, F., Mike, S., Farmer, J.: Theory for long memory in supply and demand. Physical Review E 71, 066121 (2005)
15. Toth, B., Palit, I., Lillo, F., Farmer, J.: Why is equity order flow so persistent? J. Econ. Dyn. Control. 51, 218–239 (2015)

Part II

Special Session on Agents and Mobile Devices (AM) + Learning, Agents and Formal Languages (LAFLang)

Podcast Distribution on Gwanda Using PrivHab: A Multiagent Secure Georouting Protocol

Adrián Sánchez-Carmona, Sergi Robles, and Carlos Borrego

Department of Information and Communications Engineering,
Universitat Autónoma de Barcelona (UAB)
adria.sanchez@deic.uab.cat

Abstract. We present PrivHab, a georouting protocol that improves multiagent systems itinerary decision-making. PrivHab uses the mobility habits of the nodes of the network to select an itinerary for each agent carrying a piece of data. PrivHab makes use of cryptographic techniques to make the decisions while preserving nodes' privacy. PrivHab uses a waypoint-based georouting that achieves a high performance and low overhead in rugged terrain areas that are plenty of physical obstacles. The store-carry-and-forward approach used is based on mobile agents and is designed to operate in areas that lack network infrastructure. We have evaluated PrivHab under the scope of a realistic podcast distribution application in remote rural areas. The PrivHab protocol is compared with a set of well-known delay-tolerant routing algorithms and shown to outperform them.

1 Introduction and Motivation

In 2003, the Food and Agriculture Organization of the United Nations (FAO[1]) implemented a strategic Programme entitled "Bridging the Rural Digital Divide". The programme highlighted innovative approaches to knowledge exchange that were taking advantage of new digital technologies.

E-agriculture applications, usually targeting rural areas, are very likely to deal with challenges like a sparse population, with the receivers of the information far away from each other, a bad, non-existent or expensive telephony coverage and, especially, a lack of data communication networks are the most common ones.

We propose to use PrivHab to reduce the digital divide in developing countries by distributing podcast radio programs using Mobile Agent based Delay Tolerant Networking [4]. MADTN uses mobile agents to perform a store-carry-and-forward strategy, and it is designed to operate in absence of simultaneous end-to-end paths.

[1] More information can be found on http://www.e-agriculture.org/bridging-rural-digital-divide-programme-overview

J. Bajo et al. (eds.), *Trends in Prac. Appl. of Agents, Multi-Agent Sys. and Sustainability*,
Advances in Intelligent Systems and Computing 372, DOI: 10.1007/978-3-319-19629-9_4

2 Scenario of Application

In some places, due to the region's dialect preference and the illiteracy ratios, radio broadcasting is the most important information source for farmers. It plays a key role in the economy development of the region by disseminating important agricultural information.

In Gwanda, Zimbabwe, the poor radio signal of the area leads the NGO *Practical Action*[2] to use a manpower of 60 cooperators to bring podcasts to the villagers. The cooperators, equipped with portable MP3 players and speakers, physically travel to the NGO office to obtain new podcasts that they play at their assigned villages. We aim to replace this physical distribution by a digital and automated one.

We propose to create a Delay Tolerant Network using a set of small devices that can be carried by the members of the NGO's staff or by some local villagers that collaborate with them. The deployment's cost of this network nodes should be low[3], and can be considered as an investment, since the NGO will not need to spend more resources on the podcast distribution.

Between the NGO and the local radio stations there could be barriers that nodes carrying the data can not cross, as the Mtshabezi River, and there could be some locations that are very likely to have a higher density of nodes, as the markets. Therefore, data should try to follow paths that take advantage of this knowledge. For this reason, we propose a geographical routing protocol where the sender defines a set of waypoints where the data has to pass by in order to reach its destination.

3 A Habitat-Based Itinerary

A **habitat** is defined as the area where a node is more likely to be found. It is based on the assumption of social-based routing protocols that future mobility of a node will be related to its near past mobility [3]. The heatmap (Figure 1) is an extremely accurate habitat representation.

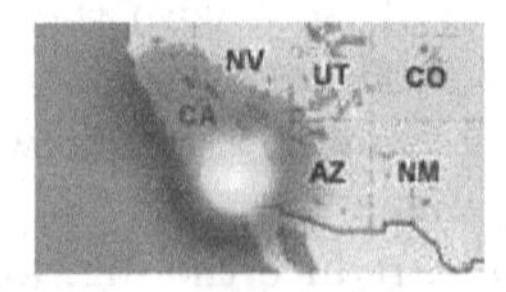

Fig. 1. Heatmap of a node. The dark red area corresponds to the area that is usually visited, and the intense yellow spot corresponds to the region where the node spends most of his time.

[2] More information about this programme at
`http://practicalaction.org/podcasting-gwanda`

[3] Small devices like Raspberry Pi can be acquired by less than 30$/unit.

However, creating and maintaining this data is a resource consuming task that does not fit well with the small devices of the proposed network. Therefore, we propose to model each nodes' habitat using a simple geometric shape. This way, nodes can automatically calculate and store their habitat consuming the minimum computational resources by using a mobile average, and they can use it to make routing decisions quickly.

3.1 Circular Model of Habitat

We model the habitat using a circle. Each habitat H is characterized by two elements: a centre point $C = (x, y)$ and a radius R. A habitat is defined by the tuple $H = (C, R)$.

Every node's habitat has to be updated in order to capture the trend of the node's mobility pattern. The update process of a habitat consists in obtaining the location of a node and adding it to his habitat's model. Nodes use the Exponentially Weighted Moving Average (EWMA) to update their previous version of the habitat, named H_{old}, with a frequency of ω updates/hour. From now on, we will refer as $L = (x_s, y_s)$ to the location of a node at the moment of the update.

Step Zero. Initialization of the Habitat: At the initialization step, H_0 is initialized with the centre point at the same coordinates of the location L_0 (node's location when the calculation starts) and $R = 0$.

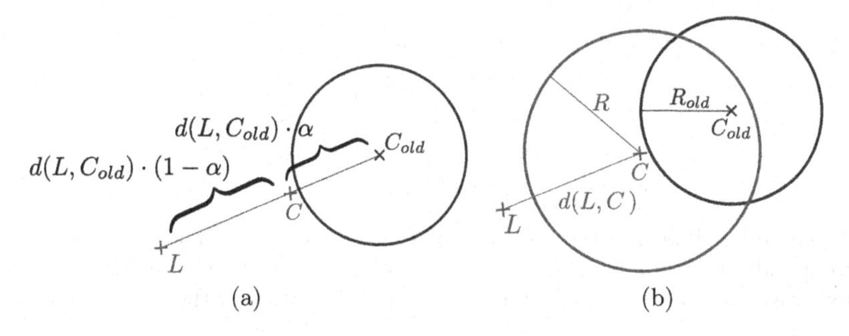

Fig. 2. Evolution of the habitat: (a) The new centre point C is calculated averaging the old centre C_{old} and the new location L; (b) The new radius R is calculated averaging the old radius R_{old} and the distance $d(L, C)$ that separates the new location L from the centre point C

First Step. Update of the Centre: The first step to updating a habitat is to update the centre. The centre point of the current habitat H is calculated by averaging using EWMA the centre point C_{old} and the current location L. The only parameter involved is α. This first step is depicted in Figure 2 (a).

$$C = L * \alpha + C_{old} * (1 - \alpha) \tag{1}$$

Second Step. Update of the Radius: After C has been calculated, the radius R is updated by averaging using EWMA the radius R_{old} of the previous habitat and $d(L, C)$, the distance between L and the centre point C. This second step is depicted in Figure 2 (b).

$$R = d(L, C) * \alpha + R_{old} * (1 - \alpha) \quad (2)$$

3.2 The Motion Common Cycle

A habitat calculated using $\alpha = \frac{2}{T\omega+1}$ models the mobility habits of a node during the last T hours. The amount of hours T a habitat models is called the common motion cycle, and it has to be known by all nodes of the network. In a mobile average, each time a location is used to update the habitat, previous locations lose weight. Concretely, in EWMA, the last $T\omega$ locations weight the 86% of the total, while previous locations weight the remaining 14%.

4 The PrivHab Protocol

The PrivHab routing algorithm compares two nodes and decides who is the best choice to carry the data towards its destination[4]. The routing algorithm chooses the nodes whose habitat's border is closer to the next waypoint, prioritizing those nodes whose habitat encloses it. If a waypoint is contained inside two different habitats, then the routing algorithm chooses the node with the smallest one.

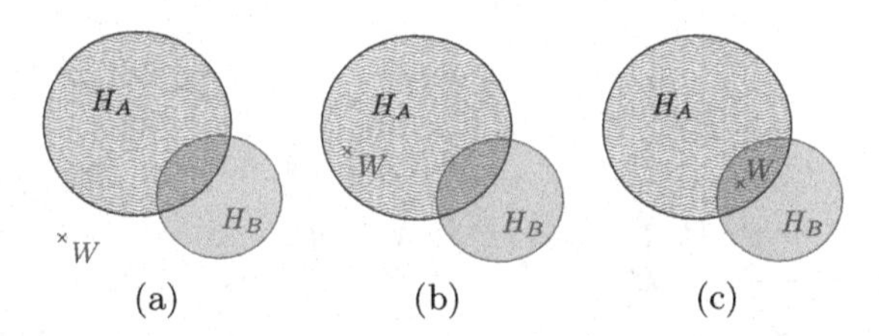

Fig. 3. Three possible situations in habitat-based routing: (a) The next waypoint is located outside the two habitats; (b) Only one of the habitats encloses the location of the next waypoint; (c) The two habitats enclose the location of the next waypoint.

Figure 3 show the different situations that can be faced. In (a) and (b) node A is chosen as the best option, because the waypoint W is closer to H_A or inside it. In (c) the best choice is B, because both habitats contain W, but H_B is smaller than H_A.

4.1 Nodes' Privacy

At [1], Boldrini *et al.* recognize that privacy is an important issue in a routing protocol. Therefore, PrivHab needs to be secure and do not reveal the habitat

[4] We assume that the approximate locations the data has to pass to reach the destination can be known or guessed by the sender.

information to the other part. For this reason, PrivHab uses the Paillier [6] additive homomorphic cryptography to protect nodes' privacy. This way, the habitats and the waypoints are operated and compared while cryptographically protected in order to avoid revealing this private information to the other parts.

4.2 Exchanged Messages

We assume that every location can be mapped to two-dimensional coordinates with a mapping known to both A, the node that carries the data, and B, a candidate neighbour. Let A's habitat be $H_A : (C_A, R_A)$. Let $W[i] : (x_{W[i]}, y_{W[i]})$ be the next waypoint. Let B's habitat be $H_B : (C_B, R_B)$. We denote $E_Y(m)$ as the Paillier additive homomorphic encryption of m using Y's public key.

1. Node A calculates $d_A = d(H_A, W[i])^2$, the square of the distance between its habitat and $W[i]$ ($d_A = 0$ if $W[i] \in H_A$ and $d_A \geq 1$ otherwise). A knows both H_A and $W[i]$, so the calculation of d_A can be performed without using homomorphic encryption.
2. Node B announces to A the centre $C_B : (x_{C_B}, y_{C_B})$ of its habitat.

   $B \rightarrow A$: $E_B(x_{C_B}), E_B(y_{C_B})$

3. Node A subtracts the coordinates of $W[i]$ to the coordinates of C. Then, A multiplies both results by the same *nonce* (a random one-use value).

$$(E_B(x_{C_B})/E_B(x_{W[i]}))^{nonce} = E_B((x_{C_B} - x_{W[i]}) \cdot nonce) \quad (3)$$

$$(E_B(y_{C_B})/E_B(y_{W[i]}))^{nonce} = E_B((y_{C_B} - y_{W[i]}) \cdot nonce) \quad (4)$$

   Following, A sends to B the results and the coordinates of $W[i]$, the distance d_A, the radius R_A, and the information B needs to calculate d_B.

$$A \rightarrow B: \begin{array}{l} E_B((x_{C_B} - x_{W[i]}) \cdot nonce), E_A(x^2_{W[i]}), E_A(R_A), E_A(2y_{W[i]}), E_A(2x_{W[i]}), \\ E_B((y_{C_B} - y_{W[i]}) \cdot nonce), E_A(y^2_{W[i]}), E_A(d_A), E_A(x_{W[i]}), E_A(y_{W[i]}) \end{array}$$

4. B decrypts the received subtractions and computes β.

$$\beta = \tan^{-1}(((y_{C_B} - y_{W[i]}) \cdot nonce)/((x_{C_B} - x_{W[i]}) \cdot nonce)) \quad (5)$$

   Node B uses β to calculate $X : (a = x_{C_B} - R_B \cdot \cos\beta, b = y_{C_B} - R_B \cdot \sin\beta)$, X is the nearest point of H_B to $W[i]$. Then, B calculates $d(H_B, W[i])^2 = d_B$, the square of the distance.

$$(E_A(a^2) + E_A(b^2))/(E_A(2x_{W[i]})^a \cdot E_A(x^2_{W[i]}) \cdot E_A(2y_{W[i]})^b \cdot E_A(y^2_{W[i]})) =$$

$$E_A(a^2 - 2ax_{W[i]} - x^2_{W[i]} + b^2 - 2by_{W[i]} - y^2_{W[i]}) =$$

$$E_A((a - x_{W[i]})^2 + (b - y_{W[i]})^2) = E_A(d_B) \quad (6)$$

Following, B calculates the point inclusion of $W[i]$ in H_B using Equation 7, the comparison of distances using Equation 8, and the comparison of radius using Equation 9. This time, three different *nonce* values are used to randomize the results. The d_A factor is used to blur[5] the point inclusion test and the comparison of radius.

$$(E_A(R_B^2) \cdot E_A(d_A))/(E_A(d_B))^{nonce} = E_A((R_B^2 + d_A - d_B) \cdot nonce) \quad (7)$$

$$(E_A(d_A))/(E_A(d_B))^{nonce} = E_A((d_A - d_B) \cdot nonce) \quad (8)$$

$$(E_A(R_A) \cdot E_A(d_A)^{R_B})/(E_A(R_B))^{nonce} = E_A((R_A + d_A \cdot R_B - R_B) \cdot nonce) \quad (9)$$

Finally, B orders the results of the two comparisons and the point inclusion test in a random way and sends it to A.

$$B \rightarrow A: \begin{array}{l} E_A((R_A + d_A \cdot R_B - R_B) \cdot nonce), E_A((d_A - d_B) \cdot nonce), \\ E_A((R_B^2 + d_A - d_B) \cdot nonce) \end{array}$$

5. Node A decrypts the three received values. B is considered a better choice if the three decrypted values are equal or greater[6] than 0.

4.3 A Multiagent System

PrivHab is executed under the MADTN framework. The agents involved in this multiagent system are listed below.

- **Habitat agent**: The agent that performs the operations described in Section 3.1 to calculate and update the habitat of the node. This agent also periodically informs the Carrier agent of the current location to track if the node had approached enough the current waypoint.
- **Interactor agent**: The agent that performs the exchange of messages described in Section 4.2. This agent informs the Carrier agent with the result when the exchange of messages has finished.
- **Carrier agent**: A proactive agent that carries the data towards its destination. After the execution of PrivHab, it makes the decision of migrating, being cloned, or staying at the current node.

[5] If $d_A > d_B$, then the best choice is B, and the result of the point inclusion test and the comparison of radius are not needed.

[6] PrivHab checks several times if an operand ρ is negative. As ρ is an element of $\mathbb{Z}_n$, to check this condition, we ensure that n is sufficiently large and that all values ρ we will use are $\rho \leq n/2$. Then, we can consider that $\rho > n/2 \iff \rho < 0$.

5 Experiments and Results

As a proof-of-concept we have deployed an implementation of the presented protocol on three Raspberry Pi boards. We have used them to measure the overhead that PrivHab adds to every transaction.

We have used our proof-of-concept implementation, using Paillier's length keys of 512, 1024 and 2048 bits, to forward 600 podcasts of sizes between 10MB and 20MB[7]. We have repeated the tests five times. We have measured the average time needed to make the calculations and to exchange all the messages. The obtained results have been incorporated to the simulations.

PrivHab execution time depends heavily on the key length used. When using keys of 512 bits, PrivHab can be executed by a low-end device in 0.57 seconds. Meaning an overhead of less than 3.48% when sending messages larger than 10MB. The execution time increases to 3.97 ± 0.03 seconds when using keys of 1024 bits. Given the average length of connectivity windows in remote village scenarios presented in [2], this overhead is acceptable. When using keys of 2048 bits, the execution time is too high ($25,031.5 \pm 69.8$ seconds).

5.1 Modelling and Simulations

The scenario we have used in all the simulations is the one presented in Section 2. We have compared the performance of PrivHab with a bench-mark of well-known routing protocols used in [5]: Prophet, Binary Spray & Wait (L=40), Epidemic and Random. We have added two routing protocols to this set: MaxProp and First Contact. All simulations have been performed using *The Opportunistic Network Simulator* (The ONE), and have been repeated twenty times using different random seeds.

The performance of all the compared protocols is presented in Figure 4. Single-copy protocols, as Random and First Contact, do not fill up the buffers. Therefore, they obtain medium delivery ratios because nodes are not forced to drop podcasts. However, their decision making is poor, and podcasts last longer on the network. For this reason, their latency is high and they produce an enormous amount of aborted relays. Flooding-based protocols, as Epidemic and Prophet, generate en enormous network overhead that fill the buffers early. Therefore, they obtain medium latencies but low delivery ratios because almost all nodes effort while forwarding podcasts is wasted, usually because the podcasts are dropped. BS&W and MaxProp perform well in terms of latency. But their performance in terms of delivery ratio is totally opposed. Binary Spray & Wait, performs poor in terms of delivery ratio because of his epidemic-style spread, while MaxProp obtains a high delivery ratio because his dropping policy based on probabilities of delivery manages to drop less messages. PrivHab takes the best decisions

[7] This is the size of an audio file with ID3 version 2.4.0, extended header, containing: MPEG ADTS, layer III, v1, 128 kbps, 44.1 kHz, stereo, with a duration between 10 and 20 minutes.

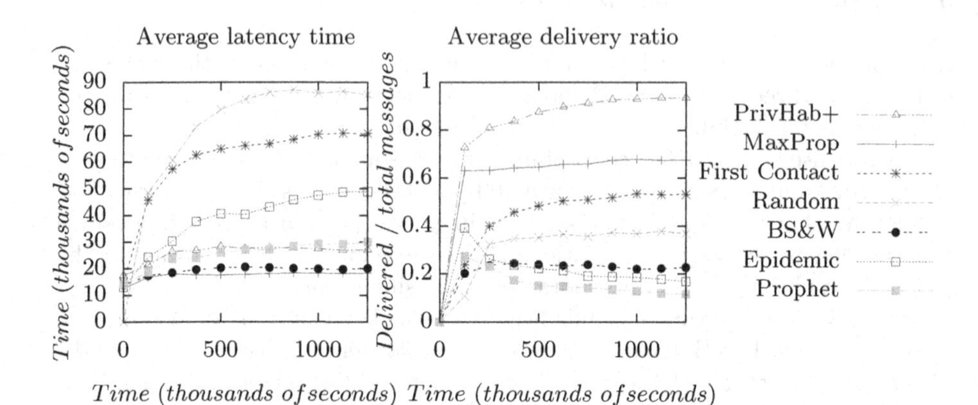

Fig. 4. Results of the simulations. Latency and delivery ratio.

because it takes into account both the pathway to the destination and the mobility patterns of the neighbours, and obtains the lowest network overhead and latency latency of the single-copy protocols because the spread is directed towards the destination.

6 Conclusions

The habitat models node's whereabouts based on the common motion cycle. It is used to decide what nodes are good choices to carry the data towards its destination. PrivHab uses homomorphic cryptography to preserve nodes' privacy.

Acknowledgment. This work has been partially funded by the Ministry of Science and Innovation of Spain, under the reference project TIN2010-15764 and by the Catalan Government under the reference project 2014SGR691.

References

1. Boldrini, C., Conti, M., Jacopini, J., Passarella, A.: Hibop: a history based routing protocol for opportunistic networks. In: IEEE International Symposium on a World of Wireless, Mobile and Multimedia Networks, WoWMoM 2007, pp. 1–12 (June 2007)
2. Grasic, S., Lindgren, A.: Revisiting a remote village scenario and its dtn routing objective. Computer Communications 48, 133–140 (2014)
3. Hui, P., Crowcroft, J., Yoneki, E.: Bubble rap: Social-based forwarding in delay-tolerant networks. IEEE Transactions on Mobile Computing, 10(11), 1576–1589 (2011)
4. Martínez-Vidal, R., Castillo-Pérez, S., Robles, S., Cordero, M., Viguria, A., Giuditta, N.: Mobile-agent based delay-tolerant network architecture for non-critical aeronautical data communications. In: Omatu, S., Neves, J., Rodriguez, J.M.C., Paz Santana, J.F., Gonzalez, S.R. (eds.) Distrib. Computing & Artificial Intelligence. AISC, vol. 217, pp. 513–520. Springer, Heidelberg (2013)

5. Musolesi, M., Mascolo, C.: Car: Context-aware adaptive routing for delay-tolerant mobile networks. IEEE Transactions on Mobile Computing 8(2), 246–260 (2009)
6. Zhong, G., Goldberg, I., Hengartner, U.: Louis, lester and pierre: Three protocols for location privacy. In: Borisov, N., Golle, P. (eds.) PET 2007. LNCS, vol. 4776, pp. 62–76. Springer, Heidelberg (2007)

Learning, Agents, and Formal Languages: Linguistic Applications of Interdisciplinary Fields

Leonor Becerra-Bonache[1] and M. Dolores Jiménez-López[2]

[1] Laboratoire Hubert Curien, Jean Monnet University,
18 rue Benoit Lauras, 42100, Saint-Etienne, France
leonor.becerra@univ-st-etienne.fr
[2] Research Group on Mathematical Linguistics,
Universitat Rovira i Virgili, Av. Catalunya 35, 43002 Tarragona, Spain
mariadolores.jimenez@urv.cat

Abstract. This paper focuses on three areas: machine learning, agent technologies and formal language theory. Our goal is to show how the interrelation among agents, learning and formal languages can contribute to the solution of a challenging problem: the explanation of how natural language is acquired and processed. Linguistic contributions of the intersection between machine learning and formal language theory –through the field of grammatical inference– are reviewed. Agent-based formal language models as colonies, grammar systems and eco-grammar systems have been applied to different natural language issues. We review the most relevant applications of these models.

1 Introduction

Nowadays, interdisciplinary research is key to make progress and increase the rate of scientific findings in different areas. There are problems that cannot be approached just by the single perspective of a specific field. Therefore, to understand better or solve these kind of problems, the collaboration of researchers from different disciplines is required.

Taking into account the relevance of interdisciplinarity, we consider the relationship among *machine learning, formal languages* and *agent technologies*:

- **Machine Learning** is one of the most active research areas within Artificial Intelligence. Its main goal is to develop techniques that allow computers to learn. Machine learning algorithms construct a model based on the inputs that they receive, and then they use that model to make predictions or decisions. Examples applications of machine learning are: spam filtering, handwriting recognition, computer vision, etc.
- **Formal Languages** was originated from mathematics (researchers as Thue, Post, Turing) and linguistics (Chomsky). Formal language theory provides mathematical tools for the description of linguistic phenomena. It was born in the middle of the 20th century as a tool for modeling and investigating the syntax of natural languages. However, very soon it developed as a new research field, separated from Linguistics, with specific problems, techniques and results and, since then, it has had an important role in the field of Computer Science.

J. Bajo et al. (eds.), *Trends in Prac. Appl. of Agents, Multi-Agent Sys. and Sustainability*,
Advances in Intelligent Systems and Computing 372, DOI: 10.1007/978-3-319-19629-9_5

- **Agent Technologies** is one of the most important areas emerged in Information Technology in the 90's. By implementing autonomous entities driven by beliefs, goals, capabilities, plans and agency properties, agent technologies capture essencial aspects of the modeled systems. The metaphor of autonomous problem solving entities cooperating and coordinating to achieve their objectives is a natural way of conceptualizing many problems. In fact, the multi-agent system literature spans a wide range of fields.

Our main goal here is to review the *linguistic* contributions of the interrelation of those three areas. This is, we want to show how the interdisciplinary relation between machine learning, formal languages and agent technologies can contribute to the solution of one of the most persistent problems in science: the explanation of how natural language is acquired and processed.

If we want to explain natural language, we need to cross traditional academic boundaries in order to solve the different problems related to this topic. We should attack the subject from various angles and methods, eventually across disciplines, forming a new method for understanding natural language. Therefore, interdisciplinarity should be an essential trait of the research in language. In this paper, we review some interdisciplinary research performed in this direction. Specifically, in section 2, the intersection between machine learning and formal languages is taken into account. This intersection constitutes a well-established research field known as *Grammatical Inference*. We review here the main linguistic applications of this field. In section 3, the interrelation between formal language theory and agent technologies is considered. We show how agent technologies can offer good solutions and alternative frameworks to classic models in the area of processing and computing languages that can be useful for the description and analysis of natural language.

2 Learning and Formal Languages

The intersection between the field of *machine learning* and the *theory of formal languages* constitutes a well-known research field called *Grammatical Inference* (GI) [13]. GI studies how grammars can be learnt from a set of data. The field of GI was originated in the 60's, mainly by the work developed by E.M. Gold [15]. Motivated by the problem of how children acquire their native language, E.M. Gold tried to investigate, from a theoretical point of view, how the ability to speak a language can be achieved in an artificial way. Since then, a big amount of research has been done by researchers coming from different scientific traditions: machine learning, formal languages, computational linguistics, pattern recognition, etc. Two main approaches can be distinguished in GI: 1) *Theoretical approaches*: researchers aim to prove efficient learnability of grammars. Most of GI researchers have been focused on this approach. Their aim is to obtain formal results, for example: formal descriptions of the target languages, formal proofs about the efficiency of a learning algorithm, etc.; and 2) *Practical approaches*: researchers aim to develop systems that learn grammars from real data. Instead of proving the learnability of grammars, researchers focus on providing empirical systems that deal with natural language data.

Despite the original linguistic motivation of the GI studies, most of the work in this field have been focused on obtaining formal results, without exploiting the linguistic relevance of the classes that have been studied and the results obtained. We will review next some practical studies in GI based on natural language data.

2.1 Practical Studies with Natural Language Data

Although most part of the work in GI are theoretical, we can also find some practical approaches in GI based on natural language data. Next we review some of the main approaches.

First of all, it is important to point out that in a GI problem, there are two different actors involved: a teacher and a learner. The teacher provides information to the learner, and the learner (or learning algorithm), from that information, must identify the underlying language. Depending on how this information is provided to the learner, we can distinguish three different GI approaches to natural language [14]: 1) *Unsupervised approach*: the teacher provides *unlabeled* examples to the learner, that is, the learner does not receive explicit information about the structure of the sentences in the target language.; *Supervised approach*: the input data consists of a set of *labeled* examples, that is, the learner receives examples of inputs paired with the corresponding correct outputs; and 3) *Semi-supervised approach*: in addition to the *labeled* examples, the learner receives *unlabeled* examples.

Like theoretical studies developed in GI, most part of GI methods for natural language have also been focused on CF grammars. Most of these natural language learning methods are based on an *unsupervised approach*, and only use *positive data* during the learning process (i.e., the learner only receives examples of sentences that belong to the target language). The most common method to evaluate these systems is by using a treebank (i.e., a linguistic corpus in which sentences are annotated with their syntactic structure, often represented in the form of a tree). This method consists in extracting from a treebank (selected as the "gold standard") a set of plain natural language sentences and giving it to the algorithm as input. Then, the GI algorithm generates structured sentences and these sentences are compared against the original structured sentences from the treebank. Different metrics can be used to do this comparison, but the most used are precision (which shows how many learned structures are correct, describing in that way the correctness of the learned grammar) and recall (which shows how many of the correct structures have been learned, giving in that way the completeness of the learned grammar). It is worth noting that one of the most used treebanks is ATIS (Air Traffic Information System), an English corpus that contains mostly questions and imperative sentences on air traffic. Examples of GI systems for natural language learning are: EMILE [1], ABL (Alignment-Based Learning algorithm) [20] and ADIOS (Automatic DIstillation Of Structure algorithm) [19]. For detailed information about these algorithms and other grammar inference methods, see [14].

In GI there has also been some efforts for taking into account more natural aspects during the language learning process. An example of it is [2,3,4]. The purpose of this work was not to learn a CF grammar, but to investigate the effect of semantics and corrections in the process of learning to understand and speak a natural language. In fact, in the early stages of children's linguistic development, semantic information seem to play an important role [10]. Taking into account that most works in GI have only been focused on syntax learning and results in GI show that language learning is hard, the following question was posed: can semantic information simplify the learning problem? In order to answer this question, a simple computational model was developed which takes into account semantics for language learning. The model was tested with ten different natural languages by using a simplified version of the Miniature Language Acquisition task (this task involves sentences that describe visual scenes), and the results show that: i) access to semantic information facilitates language learning; ii) the presence of meaning-preserving corrections has an effect of language learning, even if the learner does not treat them specially. Therefore, the results obtained with this work were not only of interest for GI, but also for the linguistic community.

Researchers in GI have also developed methods for other tasks, such as machine translation. For example, we can find several works focused on learning stochastic finite-state transducers (SFST) for machine translation [9]. A SFST involves two different alphabets (source and target alphabet), and associates probabilities to the transitions and final states. The main advantages of working with these models are the following: i) there exist efficient search algorithms for translation; ii) these techniques are less computational expensive (in general) than most pure statistical approaches; iii) they allow an easy integration with other information sources, such as acoustics models, making easier the applications of SFST to more difficult tasks, such as speech translation. It is worth noting that these methods have been successfully applied to different non-trivial tasks, such as Miniature Language Acquisition task, Traveler task, etc. For more information, the reader is refered to [9].

3 Agents and Formal Languages

In this section, we consider the interrelation between *multi-agent systems* and *formal language theory*. The first generation of formal grammars, based in rewriting, formalized classical computing models. At that time, linguistics was the central application of formal language theory and linguists were very much interested in applying formal language models to the formalization of natural language. However, from the 90s, the interest of linguistics in formal languages seems to have disappeared and formal language theorists have found innumerable applications of their theory different from linguistics. Problems related to the first generation of formal languages based on rewriting systems were the reason for that divorce between formal languages and linguistics. However, models proposed from the 90s in the area of formal language theory may solve those classic problems. Among those new models we find the *agent-based models* of formal languages that constitute an important subfield of the theory. The main advantage of those agent-based models is that they increase the power of their component grammars thanks to interaction, distribution and cooperation. *Colonies*, *grammar systems* and *eco-grammar systems* are examples of this new generation of formal languages. All these new types of formalisms have been proposed as grammatical models of agent systems. These multi-agent formal languages have been applied to natural language description and processing. In general, it can be shown that those non-standard models in formal languages can solve the classical problems related to the first generation of formal languages and can cover the whole range of linguistic disciplines, from phonology to pragmatics. In the next section we show some examples of these possible linguistic applications.

3.1 Linguistic Applications of Agent Based Formal Language Theory

Colonies are the first agent-based model we consider. Colonies as well-formalized language generating devices have been proposed in [17], and developed during the nineties in several directions. Colonies can be thought of as grammatical models of multi-agent systems motivated by Brooks' subsumption architectures [8]. They describe language classes in terms of behavior of collections of very simple, purely reactive, situated agents with emergent behavior. Colonies, as proposed originally, capture some formal aspects of systems of finite number of autonomous components capable to perform very simple reactive computing tasks each. The behaviour of the colony really emerges from interactions of its components with their symbolic environment and can considerably surpass the individual behaviours of its components. The main advantage of colonies is their generative power, the class of languages describable by colonies that make use of strictly regular components is beyond the set describable in terms of individual regular grammars.

In [7], colonies have been proposed as a tool to generate natural language by the interaction of a finite number of finite-state devices that generate finite languages. This application takes into account the idea of describing natural languages as a number of modules that interact in a nonsimple way. Colonies offer a modular theory where the various dimensions of linguistic representation may be arranged in a distributed framework and where the language of the system is the result of the interaction of those independent cooperative modules. Colonies allow us to generate infinite languages by only using grammars generating finite languages. This formal framework increases the power of regular grammars thanks to interaction. What is important here is the fact that although the generative power of colonies goes beyond the regular family of languages, the derivation process is done in a regular (finite-state) manner. Therefore, colonies may reveal as a device able of conjoining the simplicity of finite-state machines with a stronger generative power able to account for the infiniteness (context-free or more) of natural languages.

Another important agent based-model in formal languages are the so-called *grammar systems* [11]. Grammar system theory is a consolidated and active branch in the field of formal languages that provides syntactic models for describing multi-agent systems at the symbolic level, using tools from formal grammars and languages. A grammar system is a set of grammars working together, according to a specified protocol, to generate a language. Note that while in classical formal language theory *one* grammar (or automaton) works individually to generate (or recognize) *one* language, here we have *several* grammars working together in order to produce *one* language. The theory was launched in 1988 and has developed into several directions, motivated by several scientific areas.

Grammar systems may offer useful tools to account for arrangement and interaction of the various dimensions of natural language grammar. In order to define a grammar system approach for grammar architecture, a set of postulates of linguistic theory must be followed [18]: 1) we take a grammar to be a set of subgrammars called modules; 2) each of these modules is a grammar of an independent level of linguistic representation; these modules are not hierarchically related to one another; 3) a module need not wait for the output of another to do its work, but has the power to generate (analyze) an infinite set of representations quite independently of what is going on in any of the other components; 4) each component is a self-contained system, with its own independent set of rules, principles and basic vocabulary; 5) the lexicon plays a special, transmodular role in the theory. In order to capture every feature of the above list, a new variant of grammar systems has been introduced: *Linguistic Grammar Systems* (LGS) [16]. LGS offer an example of the possible application of formal languages to linguistics. For formal definitions of LGS the reader can see [16].

Very relevant in the area of multi-agent models of formal languages are the so-called *eco-grammar systems* [12]. Eco-grammar systems provide a syntactical framework for eco-systems, this is, for communities of evolving agents and their interrelated environment. An eco-grammar system is defined as a multi-agent system where different components, apart from interacting among themselves, interact with a special component called 'environment'. Within an eco-grammar system we can distinguish two types of components *environment* and *agents*. Both are represented at any moment by a string of symbols that identifies the current state of the component. These strings change according to sets of evolution rules. Interaction among agents and environment is carried out through agents' actions performed on the environmental state by the application of some productions from the set of action rules of agents.

Eco-grammar systems can model the structure of dialogue and account for the evolution of language. According to the idea that dialogue can be understood as the sustained production of mutually-dependent acts, constructed by two or more agents each monitoring and building on the actions of the other, eco-grammar systems can describe dialogue as a sequence of *acts* performed by two or more agents in a common environment. An example of this application is presented in [6]. In this paper, a formal model of dialogue based on eco-grammar systems is introduced: *Conversational Grammar Systems* (CGS). CGS present some advantages to account for dialogue: a) the generation process is highly *modularized* by a distributed system of contributing agents; b) it is *contextualized*, linguistic agents re-define their capabilities for acting according to context conditions given by mappings; c) and *emergent*, it emerges from current competence of the collection of active agents.

4 Conclusion

We have considered three different areas: agent systems, machine learning and formal languages. Each of those fields has intrisically good features that enable them to cope with many real-world problems. The formal apparatus of agent technology provides a powerful and useful set of structures and processes for designing and building complex applications. Machine learning is one of the core fields of Artificial Intelligence since the ability to learn is one of the most fundamental attributes of intelligent behavior. And finally, formal language theory provides the flexibility and the abstraction necessary in order to be applied to fields such as linguistics, economic modeling, developmental biology, cryptography, sociology, etc. Therefore, multi-agent systems, machine learning and formal language theory provide flexible and useful tools that can be used in different research areas due to their versatility. In this paper, we have tried to show that the individual power of those systems may be increased if they collaborate among them.

In this work, we have focused on two possible intersections: formal languages and learning; and, agents and formal languages. Our main objective here has been to show the contributions of those intersections to the area of natural language processing. The interaction between researchers in those three topics can provide good techniques and methods for improving our knowledge about how languages are processed.

Language is one of the most challenging issues that remain to be explained. Natural language is a hard problem not only for linguistics that has not yet provided universal accepted theories about how language is acquired and processed, but also for computer science that up to now has implemented natural language processing systems that are far from being satisfactory. As a complex system, the explanation, formal modelling and simulation of language present important difficulties. If we deal with language, we need to connect and integrate several academic disciplines in order to find a solution. In this interdisciplinary environment, formal languages, machine learning and agents systems can collaborate –as shown in this paper– in the description, explanation and processing of language.

References

1. Adriaans, P.: Language learning from a categorial perspective. PhD thesis, University of Amsterdam (1992)
2. Angluin, D., Becerra-Bonache, L.: Learning meaning before syntax. In: Clark, A., Coste, F., Miclet, L. (eds.) ICGI 2008. LNCS (LNAI), vol. 5278, pp. 1–14. Springer, Heidelberg (2008)
3. Angluin, D., Becerra-Bonache, L.: Effects of meaning-preserving corrections on language learning. In: CoNLL 2011, pp. 97–105 (2011)
4. Angluin, D., Becerra-Bonache, L.: A Model of semantics and corrections in language learning. Technical Report, Yale University, 1–45 (2010)
5. Becerra-Bonache, L., Case, J., Jain, S., Stephan, F.: Iterative learning of simple external contextual languages. Theoretical Computer Science 411, 2741–2756 (2010)
6. Bel-Enguix, G., Jiménez-López, M.D.: Modelling dialogue as inter-action. International Journal of Speech Technology 11(3/4), 209–221 (2008)
7. Bel-Enguix, G., Jiménez-López, M.D., Martín-Vide, C.: Using finite-state methods for getting infinite languages: A preview. Romanian Journal of Information, Science and Technology 12(2), 125–137 (2009)
8. Brooks, R.A.: Elephants don't play chess. Robotics and Autonomous Systems 6, 3–15 (1990)
9. Casacuberta, F., Vidal, E.: Learning finite-state models for machine translation. Machine Learning 66(1), 69–91 (2007)
10. Chouinard, M.M., Clark, E.V.: Adult reformulations of child errors as negative evidence. Journal of Child Language 30, 637–669 (2003)
11. Csuhaj-Varjú, E., Dassow, J., Kelemen, J., Păun, G.: Grammar systems: A grammatical approach to distribution and cooperation. Gordon and Breach, London (1994)
12. Csuhaj-Varjú, E., Kelemen, J., Kelemenová, A., Păun, G.: Eco-grammar systems: A grammatical framework for life-like interactions. Artificial Life 3(1), 1–28 (1996)
13. de la Higuera, C.: Grammatical inference: Learning automata and grammars. Cambridge University Press, Cambridge (2010)

14. D'Ulizia, A., Ferri, F., Grifoni, P.: A survey of grammatical inference methods for natural language learning. Artificial Intelligence Review 36(1), 1–27 (2011)
15. Gold, E.M.: Language identification in the limit. Information and Control 10, 447–474 (1967)
16. Jiménez-López, M.D.: A grammar systems approach to natural language grammar. Linguistics and Philosophy 29, 419–454 (2006)
17. Kelemen, J., Kelemenová, A.: A grammar-theoretic treatment of multiagent systems. Cybernetics and Systems 23, 621–633 (1992)
18. Sadock, J.M.: Autolexical syntax. A theory of parallel grammatical representations. University of Chicago Press, Chicago (1991)
19. Solan, Z., Horn, D., Ruppin, E., Edelman, S.: Unsupervised learning of natural languages. PNAS 102(33), 11629–11634 (2005)
20. van Zaanen, M.: Bootstrapping structure into language: alignment-based learning. PhD thesis, University of Leeds (2001)

Assessing an Ontology for the Representation of Clinical Protocols in Decision Support Systems

Tiago Oliveira, Paulo Novais, and José Neves

Algoritmi Research Centre/Department of Informatics,
University of Minho, Braga, Portugal
{toliveira,pjon,jneves}@di.uminho.pt

Abstract. In order to assess the expressiveness of the CompGuide ontology for Clinical Practice Guidelines, a study was conducted with fourteen students of the Integrated Masters in Biomedical Engineering from the University of Minho in Portugal to whom it was proposed the representation of multiple guidelines according to the ontology. They were then asked to evaluate the ontology through a questionnaire and written reports. Although the results seem promising, there is the need for significant improvements mainly in: the representation of medication prescriptions, the tasks used to retrieve information from the patient, the diversity of actions offered by the ontology, the expressiveness of conditions regarding the state of a patient, and temporal constraints.

Keywords: Clinical Protocols, Ontologies, Clinical Decision Support.

1 Introduction

There are various ways of expressing medical knowledge in Clinical Decision Support Systems (CDSSs) [5], but, among them, decision trees, probabilistic models, and task-network models (TNM) are arguably the most popular [8]. Despite the obvious advantages of each one, task-network models are still preferred over the others because of their representation of clinical guideline knowledge in hierarchical structures containing networks of clinical actions that unfold over time. The main TNMs include Asbru [9], PROforma [2], GLIF3 [1], SAGE [12], and GLARE [11], among others. The first one, Asbru [9], focuses on temporal parameters and offers constructs to define starting points, durations, and ending points of tasks. In addition, it allows the specification of intentions for actions and prescriptions, as well as the expected outcomes. These time-oriented actions, conditions and intentions are represented as patterns which assume the hierarchical structure of plans and sub-plans. As for PROforma [2], it follows a structure somewhat similar to Asbru in the sense that it also resorts to the representation of guidelines as plans. There is a root task to which every plan in a guideline belongs.In turn, a plan has any number of instances of other tasks, from actions to decisions. Its focus is on argumentation in favor or against a decision. GLIF3 [1] was the first model to place its emphasis on the use of standards. In addition to using the task model just as the other models do, it makes use

J. Bajo et al. (eds.), *Trends in Prac. Appl. of Agents, Multi-Agent Sys. and Sustainability*,
Advances in Intelligent Systems and Computing 372, DOI: 10.1007/978-3-319-19629-9_6

of terminologies to avoid semantic ambiguity in the definition of clinical terms, and employs the HL7 Reference Information Model (RIM) to ensure that other systems can communicate with a system using GLIF3. As for SAGE [12], a direct evolution of GLIF3, it is considered one of the most complete approaches to Computer-Interpretable Guidelines (CIGs). This model places a high importance on the notion of context. The context coordinates the activation of guideline-based decision support. As such, it has constructs that allow the definition of the conditions in which medical practice takes place, whether they are related to the organizational setting and roles, the patient characteristics, or necessary resources. The procedural guideline logic is represented in an activity graph constrained by the already-mentioned context variables. The GuideLine Acquisition, Representation and Execution (GLARE) [11] model is specialized in the treatment of repeated (periodic) events, which play a major role in clinical therapies. There are other important models featured in comprehensive reviews such as [8] and [6]. However, there is no standard computer-interpretable representation for CPGs, and many of the existing are criticized for lack of expressiveness.

Ontologies are explicit representations of the concepts from a domain. They are the basic construction units of the semantic web and their objective is to allow applications to process the content of information, rather than just presenting it. The Web Ontology Language (OWL) [3] is a standard proposed by the World Wide Web Consortium (W3C) designed for facilitating machine interpretation. Ontologies have not been widely explored in the representation of the procedural logic in clinical protocols, yet, they provide an ideal support for this knowledge. Thus, the objective is exploring OWL as the underlying language for CIGs and use it to develop a CIG ontology with the intention of building a sufficiently expressive representation that would be capable of accommodating knowledge from different types of guidelines.

This work presents a preliminary study designed to assess the CompGuide ontology for clinical protocols. As such, the organization of the paper is as follows. Section two provides a brief description of the ontology with its main primitive classes and properties. Section three describes the materials and methods for the study. Section four presents the results and their discussion. Finally, in section five conclusions and future work considerations are provided.

2 Developed Ontology

The CompGuide ontology was initially presented in [7]. The ontology provides a task network model representation for clinical guidelines in OWL. In order to fulfill that purpose, it follows a logic in which complex information elements are represented as individuals with multiple object properties connecting them to other individuals, and simple information that cannot be further decomposed is represented using data properties. However, simple information that is reusable and will most likely be needed across different guidelines is represented as class individuals as well. In this regard the representation is similar to a linked list of procedures. As such, a CPG is represented as an instance of the *Clinical-PracticeGuideline* class. Individuals from this class have a set of data and object properties that enable the representation of descriptive and administrative guideline information such as the name of the guideline, its general description, date of creation and last update, version, clinical specialty, category, intended users, and target population. An example of the initial definition of a guideline is given in Fig. 1.

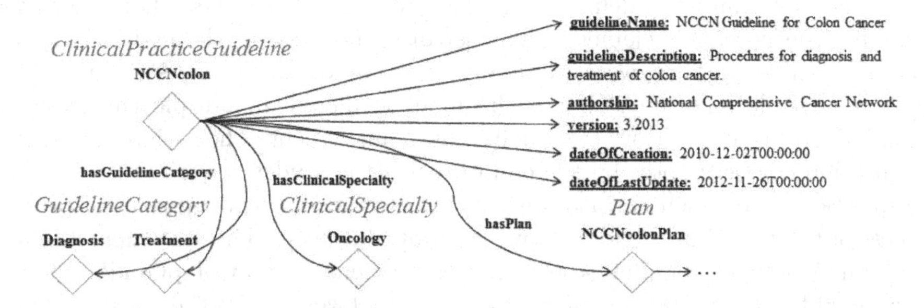

Fig. 1. Initial definition of a National Comprehensive Cancer Network guideline for the treatment of colon cancer in the CompGuide ontology.

Every guideline is connected to an individual of the class *Plan* which, in turn, is connected to other individuals that represent basic tasks. The procedural logic and workflow of clinical tasks is represented using three basic classes: *Action*, *Decision* and *Question*. The objective here is to create a recommendation plan that contains references to specific types os tasks. The *Action* class expresses a procedure that should be carried out by a health care professional. There are several subtypes of actions in the ontology that specify their nature with more detail. The *Decision* class is used to make assertions about the state of the patient, to infer new information from the existing one. The most obvious example of such a task is clinical diagnosis. The *Question* task is used to get information about the symptoms of a patient, to register information from the observations of the physician, and to store results from clinical exams. This type of task gathers all the information necessary for the execution of the clinical algorithm. Through object properties, it is possible to define the sequence of execution of tasks or if they should be executed simultaneously or concurrently.

3 Materials and Methods

The objective of the preliminary study was to assess the expressivity of the CompGuide ontology in four important aspects of CPGs, namely the representation of administrative information, the construction of workflow procedures, the definition of temporal constraints, and the definition of clinical constraints.These are considered the fundamental aspects of CIG representation and the pre-requisites of a good CIG model [8]. For that purpose, 14 students from the fourth year of the Integrated Masters in Biomedical Engineering, branch in Medical Informatics, from the University of Minho, in Braga, Portugal, aged between 22 and 23 years old, were selected. They had no prior knowlegde of OWL, and received training in both OWL and Protégé [10] for a total of six hours distributed by three two-hour sessions. After the training, they were taught about the structure, classes, and properties of CompGuide in a two-hour session.

Then, the students were asked to do an assignment which consisted in the representation of a CPG in the referred ontology using Protégé. They were handed a .owl file containing the definition of the ontology which they should fill in by adding the necessary elements. The set of CPGs used in the assignment is showed in Table 1. They were randomly distributed among the students. As much as possible, one tried that each guideline included multiple categories, namely diagnosis, evaluation, treatment, and management. The assignment had the duration of one month, by the end of which the students were asked to fill in a questionnaire which consisted of sixteen statements regarding the expressiveness of the model in the four above-mentioned aspects. The statements used in the questionnaire complete the general statement: "The CompGuide ontology allowed the representation of:". Statements 1-9 were about the construction of workflow procedures, statements 10-12 were related with the definition of clinical constraints, statements 13-15 were devised to assess the definition of temporal constraints, and, finally, statement 16 was about the representation of administrative information. The set of statements can be consulted in Figure 2. The answers were provided in a five point Likert rating scale [4] (*1-strongly disagree*, *2-disagree*, *3-neutral*, *4-agree*, *5-strongly agree*). It was also asked that the students handed a ten-page report describing their principal difficulties and observations while performing the task.

Table 1. List of the guidelines that were used in the study, featuring their name, organization and the number of people assigned to their representation.

Clinical Practice Guideline	Organization	People Assigned
Clinical Practice Guidelines in Oncology - Colon Cancer	National Comprehensive Cancer Network	2
Clinical Practice Guidelines in Oncology - Rectal Cancer	National Comprehensive Cancer Network	2
Clinical Pratice Guidelines in Oncology - Distress	National Comprehensive Cancer Network	2
Clinical Practice Guidelines in Oncology - Palliative Care	National Comprehensive Cancer Network	2
Detection,Evaluation,and Treatment of High Blood Cholesterol in Adults	National Heart Lung and Blood Institute	1
Diagnosing and Managing Asthma	National Heart Lung and Blood Institute	1
Diagnosis, Evaluation and Management of von Willebrand Disease	National Heart Lung and Blood Institute	1
Diagnosis and Treatment of Respiratory Illness in Children and Adults	Institute for Clinical Systems Improvement	1
Diagnosis and Management of Diabetes	Institute for Clinical Systems Improvement	1
Diagnosis and Treatment of Ischemic Stroke	Institute for Clinical Systems Improvement	1

The process resulted in the diverging stacked bar chart in Figure 2. The chart presents the total percentage of agreement (calculated as *agree + strongly agree*), the total percentage of disagreement (calculated as *disagree + strongly disagree*), and the percentage of participants who were neutral (equal to the percentage of the *neutral* category), for each statement, in order to show the central tendency in each item.

4 Results and Discussion

By consulting the chart of Figure 2, and specifically items 1 to 9 which refer to the representation of different procedures and tasks in a workflow, it is possible to verify that, for each item in this group, the level of agreement is at least equal or above 50%. Indeed, the item about medication prescriptions (item 1) is the one that has the lowest agreement, the highest percentage in the *neutral* category (43%), and the only one that has percentage in the *strongly disagree* category (7%). This is indicative that the representation of medication prescriptions may have issues. In fact, in the reports the participants mentioned that the representation of medication prescriptions was impractical at times. In the ontology, a medication has to be defined as the subtype of an *Action* individual, and one *Action* can only have one prescription. However, in several of the represented guidelines what were perceived as single actions included the administration of more than one drug, requiring the representation of medication schemes as several parallel actions instead of a single action with a clear objective. Another criticism to the representation of medication prescriptions was that it was mandatory to define an active ingredient, dosage, pharmaceutical form, and posology for a drug, but in certain guidelines these elements were not

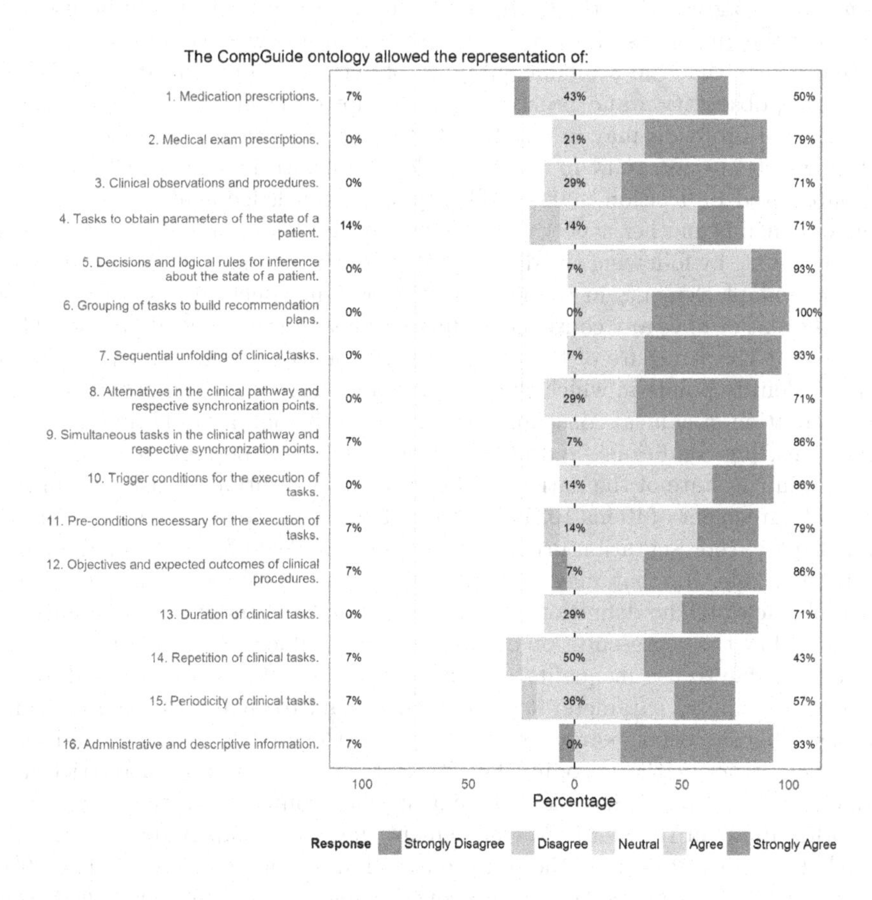

Fig. 2. Diverging stacked bar chart showing the results of the questionnaire to assess the expressiveness of the CompGuide ontology

available. Both items 1 and 2 seem to correspond to the requirements of guideline representation as they have high percentages of agreement. Item 4 also has a high percentage of agreement, but it is, among the nine items, the one that has the highest percentage of disagreement (14%). This may be due to some limitations of the *Question* class such as the absence of a description data property where it would be possible to provide a detailed description of the information that the task aims to obtain. The participants considered that the way in which the ontology is designed allows the representation of series of questions, decisions and actions, which mimics the organization of the algorithms of clinical protocols. This is evident in the high levels of agreement of items 5 to 9. Overall, the organization of the procedural logic of the guideline and the grouping of tasks in plans was considered to be advantageous, mainly because this helps the delimitation of different diagnoses, treatments, and realities. The item that refers to this grouping of tasks, item 6, has an agreement of 100%. As a whole, and given the topics presented in the questionnaire, it can be considered that the participants widely agreed that the CompGuide ontology could effectively be used to represent the guidelines in question. Nonetheless, there were concerns expressed in the reports that the available subtypes of actions (namely medication, clinical exam, observation and procedure) would not cover all the possible actions that clinical protocols may have. Many CPGs have knowledge encoded as index tables which are necessary in order to calculate health indexes which, in turn, are later used in decision making. This type of knowledge could not be represented, which is another aspect to improve. On the other hand, the participants reported that, by following the design pattern of the ontology, they were able to find redundant elements in the guideline algorithms which did not trigger any kind of event or have any consequence further ahead in the clinical process. This means that the structure of the ontology is at the same time intuitive and can help to identify points in which the integrity of guidelines are compromised. The representation of clinical constraints is central to the ontology. Through trigger conditions, pre-conditions, and expected outcomes it is possible to respond to changes in the state of the patient and control the execution of tasks. From the levels of agreement of items 10, 11, and 12, of which the lowest is 79%, it can be said that the representation primitives for these elements fulfilled, for the most part, their role. As that may be, the participants mentioned that there were some obstacles to the definition of conditions. One of them was that conditions did not allow the representation of intervals for a value of a clinical parameter. It was possible to use inequality constraints, but to define an upper and lower bound for a clinical parameter it would be necessary to create two separate conditions. This situation requires extra work from the guideline encoder and may introduce errors in the encoding.The items referring to temporal restrictions, namely items 13 to 15, have low agreement when compared to the majority of the other items in the chart. The agreement that the CompGuide ontology allowed the representation of the duration of clinical constraints was 71%. 29% of the participants answered in the neutral category. As a matter of fact, the participants observed that, while it was possible to define how long a task should last, the expressive power of the ontology was limited in this regard. It was not possible to define intervals of duration for tasks with minimum and maximum values. However, this type of information element occurred very often in the guidelines. Meanwhile, item 14, which concerns the repetition of clinical tasks, gathered only 43% of agreement, and 50% of the participants answered in the *neutral* category. Despite recognizing the usefulness of the ontology element that enables the definition of the number of times that a task should be executed, the participants believed that a crucial element was missing, and that was conditional repetitions, i.e., the possibility of stating that a task should be repeated if a the state of a patient does not improve. This was also an observation made within the scope of item 15.Finally, item 16 got 93% of agreement, which seems to convey that the ontology elements responsible for the representation of administrative information such as authoring, name of the guideline, its general description, date of creation, and so forth, are fulfilling their target function.

On a final note, representation formats such as CompGuide have to also be capable of representing situations in which the decision is left to the health care professional. Occasionally, the elements for a decision may not be all present in the description provided by a CPG, and in such situations guidelines may recommend that health care professionals follow their best judgment according to the available evidence.

5 Conclusions and Future Work

Although there was no access to an entire statistical population of interest, given the time-consuming nature of the survey, the study still provides useful hints for the development of the CompGuide ontology. Essentially, one may consider that the guidelines used in the survey were accurately represented according to the ontology, despite the need for certain adaptations, which did not affect the logic of the clinical process represented in CPGs. Nonetheless, there is a need for significant improvements, mainly in: the representation of medication prescriptions, the tasks used to retrieve information from the patient, the diversity of actions offered by the ontology, the expressiveness of conditions regarding the state of a patient, and temporal constraints as a whole.

These promising results may be due, in part, to the nature of the sample used in the survey. Since all of the participants were students of medical informatics, one can say that they have a clearer understanding than most about the role played by technology as a support for medical knowledge, and, although they were not familiar with OWL or Protégé, they already knew about similar models and principles. Moreover, the study should have included a broader range of CPGs in terms of origin and clinical specialty in order to expose the participants to a wider diversity of clinical situations. In future surveys, and once the issues identified are addressed, these aspects should be taken into consideration. Future work also includes completing the ontology proposed in this work with the international standards and data models proposed by HL7.

Acknowledgements. This work is part-funded by ERDF - European Regional Development Fund through the COMPETE Programme (operational programme for competitiveness) and by National Funds through the FCT Fundação para a Ciência e a Tecnologia (Portuguese Foundation for Science and Technology) within project FCOMP-01-0124-FEDER-028980 (PTDC/EEI-SII/1386/2012). The work of Tiago Oliveira is supported a FCT grant with the reference SFRH/BD/85291/ 2012.

References

1. Boxwala, A.A., Peleg, M., Tu, S., Ogunyemi, O., Zeng, Q.T., Wang, D., Patel, V.L., Greenes, R.A., Shortliffe, E.H.: GLIF3: a representation format for sharable computer-interpretable clinical practice guidelines. Journal of Biomedical Informatics 37(3), 147–161 (2004)
2. Fox, J., Ma, R.T.: Decision Support for Health Care: the PROforma Evidence Base. Informatics in Primary Care 14(1), 49–54 (2006)
3. Group, O.W.: OWL 2 Web Ontology Language Document Overview. Tech. Rep. W3C (October 2009), http://www.w3.org/TR/owl2-overview/
4. Jamieson, S.: Likert scales: how to (ab)use them. Medical Education 38(12), 1217–1218 (2004), http://dx.doi.org/10.1111/j.1365-2929.2004.02012.x
5. Musen, M., Shahar, Y., Shortliffe, E.: Clinical decision-support systems. In: Shortliffe, E., Cimino, J. (eds.) Biomedical Informatics. Health Informatics, pp. 698–736. Springer, Heidelberg (2006)
6. Oliveira, T., Novais, P.: Development and implementation of clinical guidelines: An artificial intelligence perspective. Artificial Intelligence Review 42(4), 999–1027 (2014)
7. Oliveira, T., Novais, P., Neves, J.: Representation of Clinical Practice Guideline Components in OWL. In: Pérez, J.B., Hermoso, R., Moreno, M.N., Rodríguez, J.M.C., Hirsch, B., Mathieu, P., Campbell, A., Suarez-Figueroa, M.C., Ortega, A., Adam, E., Navarro, E. (eds.) Trends in Pract. Appl. of Agents & Multiagent Syst. AISC, vol. 221, pp. 77–85. Springer, Heidelberg (2013)
8. Peleg, M.: Computer-interpretable clinical guidelines: A methodological review. Journal of Biomedical Informatics 46(4), 744–763 (2013)
9. Seyfang, A., Miksch, S., Marcos, M.: Combining diagnosis and treatment using ASBRU. International Journal of Medical Informatics 68(1-3), 49–57 (2002)
10. Stanford Center for Biomedical Informatics Research: Protege Wiki (2014), http://protegewiki.stanford.edu/wiki/Main_Page
11. Terenziani, P., Montani, S., Bottrighi, A., Torchio, M., Molino, G., Correndo, G.: The GLARE approach to clinical guidelines: main features. Studies in Health Technology and Informatics 101(3), 162–166 (2004)
12. Tu, S.W., Campbell, J.R., Glasgow, J., Nyman, M.A., McClure, R., McClay, J., Parker, C., Hrabak, K.M., Berg, D., Weida, T.: Others: The SAGE Guideline Model: achievements and overview. Journal of the American Medical Informatics Association 14(5), 589 (2007)

An Agent-Based Social Simulation Platform with 3D Representation for Labor Integration of Disabled People

Alberto Barriuso[1], Fernando De la Prieta[1], and Tiancheng Li[2]

[1] Department of Computer Science and Automation Control. University of Salamanca, Plaza de los caidos s/n, 37077, Salamanca
{albarriuso,fer}@usal.es
[2] School of Mechanical Engineering. Northwestern Polytechnical University, Xi'an, 710072, P.R. China
t.c.li@mail.nwpu.edu.cn

Abstract. In this paper, we present the design and implementation of a novel agent-based platform with 3D representations for labor integration. In order to evaluate and validate the proposed platform, a case study in a real workplace environment is modeled, with the purpose of identifying architectural barriers that hinder the accessibility and the processes of the workplace environment, allowing us to improve the accessibility of the environment to disabled people. Thereby, through the incorporation of novel simulation technics and mechanisms, it is possible to find areas that may be improved.

Keywords: Multi-agent system, agent-based simulation, 3D representation, jade.

1 Introduction

According to studies carried out by the United Nations (UN) [26], around 15% of the world population suffers some kind of disability, either physical or psychological, which hinders or impede them to carry out their regular daily routines in all the scopes (work, education, personal, etc.). Inasmuch as this collective faces clear disadvantages in their dairy routines to a greater or lesser extent, it is necessary to undertake all possible efforts to minimize these difficulties.

The integration of disabled people among the working environment is one of the mayor challenges in this regard, since it does not just allow the self-sufficiency of disabled people, but it also facilitates their social integration, and their self-esteem improvement. However, this is a large challenge. Speaking in quantitative terms, the *International Labour Organization* (ILO) points that there are around 386 million working-age people who suffer some kind of disability. This rate reaches values around the 80% in some countries [17].

There is no doubt that the current society is concerned about this challenge. In addition, there is a growing preoccupation from the different governments, public organizations and private entities. In this regard, since the declaration of the *Standard Rules*

J. Bajo et al. (eds.), *Trends in Prac. Appl. of Agents, Multi-Agent Sys. and Sustainability*,
Advances in Intelligent Systems and Computing 372, DOI: 10.1007/978-3-319-19629-9_7

on the Equalization of Opportunities for Persons with Disabilities by the UN [25], different national, regional and local governmental organizations have developed specific regulatory frameworks that have improved the everyday lives of people with disabilities, where we can highlight, at European level, the *Charter of Fundamental Rights of the European Union* [4], the *Treaty on the Functioning of the European Union* [7], the *Council Directive 2000/78/EC* [14] and the *European Disability Strategy 2010-2020* [6]. These initiatives go in hand with a growing academic attention, which has allowed the development of systems, techniques, models and methodologies oriented to improve the accessibility [18]. So, as example, we can highlight the work proposed by [9] that presents a software user-friendly man machine interface for accessing AT software in cloud computing or [20], where an obstacle avoidance system and a Fuzzy Logic Controller (FLC) adapted to a wheelchair is presented.

Beyond the development of these artifacts, one of the most promising study fields is social simulation, which in general terms, is applied to obtain a higher knowledge of a certain phenomenon which is being studied [5]. In the framework of the disabled people occupational integration, simulations can allow discovering in advance the problems which workers with disabilities will find before this problems occur. Within the scope of this research work, a new multi-agent system (MAS) based simulation platform is proposed. This platform makes use of 3D models of the workplace in order to perform accurate simulations, allowing at the same time to collect relevant and high quality data. Through the use of the proposed platform, it is possible to investigate in new mechanisms oriented to the decision making regarding the workflows present in the environment, the architectonical barriers detection, etc. This will allow the improvement of the workplace environment accessibility.

The rest of this paper is organized as follows: section 2 provides a brief introduction to the *Agent-Based Social Simulation* (ABSS) theory, Section 3 defines the proposed system, Section 4 evaluates the proposed system through a case of study and we conclude and introduce future research topics in Section 5.

2 Agent Based Social Simulation

ABSS is one of the most representative techniques which are used to complex inquiries where a large number of active and heterogeneous objects are present. These objects can be humans [21], business units [16], functional or nonfunctional objects [27], animals [23], etc. The phenomenon to be simulated is a set of sequences or events in a system (natural or artificial), which can exist (or not) in the real (or artificial) world and can be configured in the simulation model. These tasks are interrelated, since they finally are time or order dependent within the events which take place in the system. In this way, the simulation model allows the implementation of different and specific functionalities for each kind of target, applying different degrees of freedom. The complexity lies in the fact that this kind of models allow the simulation of complex and changing events, so the use of intelligent agent based techniques is appropriate, arising in this direction the ABSS concept [10]. This technique is focused on the social phenomenon simulation, using MAS models. Therefore ABSS is a combination of social sciences, agent-based computation and computational simulation.

The use of this techniques is specially indicated when it is necessary to capture different tasks, elements, objects, persons in dynamic complex environments, as long as they can be implemented without having a deep knowledge about the global interdependencies. Moreover, it also allows facilities on changing models, since it is not necessary to make local changes, but global [2]. Besides, the benefits of agent based computing for computer simulation include various methods for evaluating MAS or for training future users of the system [11].

The use of this models and tools in ABSS allows modeling a great variety of tasks and environments. In fact, it has been applied in previous works, obtaining good results [13], [3]. However, it has been observed that there are certain difficulties when modeling the physical environment which we intend to simulate [19]. This feature is especially relevant when modeling simulations oriented to the disabled people workplace integration, since a high problem percentage of the problems to be solved are related with the structure of buildings, or accessibility barriers (steps, ramps, lifters, doors, corridors, etc.).

3 Agent-Based Platform for Simulating Working Environments

In this section the proposed platform is presented. One of the main aims to be achieved with its design and development has been the capability of modeling different environments in a flexible and dynamic way. For that reason, a novel combination of virtual organization (OV) based MAS has been used [15], allowing to model the human organization and a 3D environment representation where simulations can be carried out. A representation of the proposed simulation architecture is illustrated in Fig. 1.

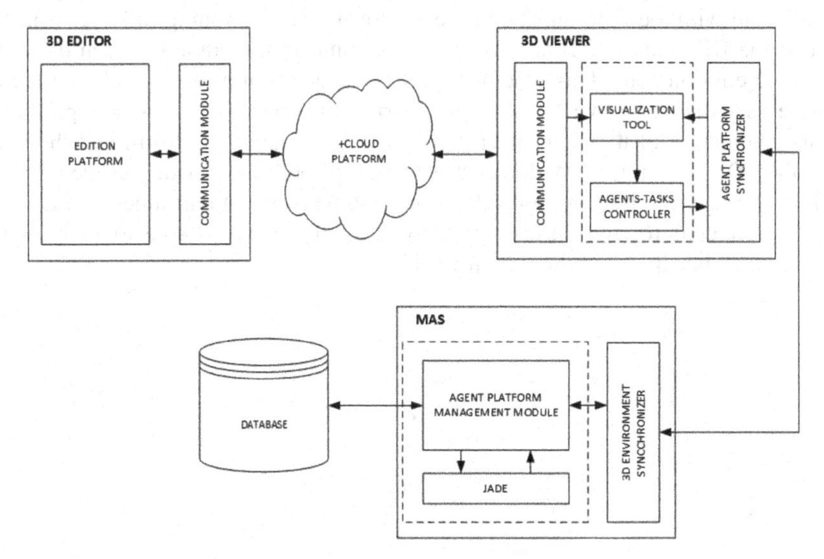

Fig. 1. Agent-based simulation platform

The combination of both strategies is a key factor to achieve the main objective which we pretend to simulate. On one hand, agents can implement different tasks, objectives, purposes, etc.; and in the other hand, the 3D environment allows modeling physically the workplace environments. Thereby, this approach allows carrying out simulations using the workplace environment itself, as the visualization of their results.

3.1 Components of the Proposed Architecture

On the one hand, **MAS** consists of intelligent agents which represent and allows to simulate the desired model. The MAS is based on VO [24], allowing the representation of a structure, roles, as well as a wide range of rules which will schedule the interactions among agents.

The MAS has been developed by means of JADE agent platform [1]. In this respect, JADE does not allow the implementation of this capabilities, so it has been necessary to include a higher layer, which is responsible for the MAS management from an organizational point of view. This layer, allows us to provide additional self-organizational capabilities to the platform, including: (i) the capacity of grouping the MAS as a VO, enabling the agents to adopt different topologies [12]; (ii) definition of the tasks attending to the role which the agent assumes inside the organization, as well as the definition of unique characteristics to each agent;(iii) the ability to define a set of rules to regulate the interactions among agents has been implemented. Furthermore, this software layer which is included over JADE also allows to define scheduling and task-allocating oriented agents within the simulation. Finally, this layer also allows the communication within the 3D environment, so it can notify all the changes that take place bidirectionally between the agent platform and the tridimensional environment.

Fig. 2 shows in detail the structure of each one of the agents, consisting of three main components: (i) the communication module, which will allow the agents to communicate with the other agents in the platform, (ii) a reasoning module, which is based on the BDI model [22] and (iii) the communication module between the agent and the 3D environment. This module is in turn made up of a sensor and an effector. It is necessary that these two elements are distributed, being one component present in the implementation of the agent itself (as a part of the agent platform) and the other one in the 3D environment. With this distribution of both the sensor and the effector, we allow the agent to be conscious about the tridimensional environment state which has been rendered for the current simulation, so the agent is able to monitor the changes which take place in the virtual world.

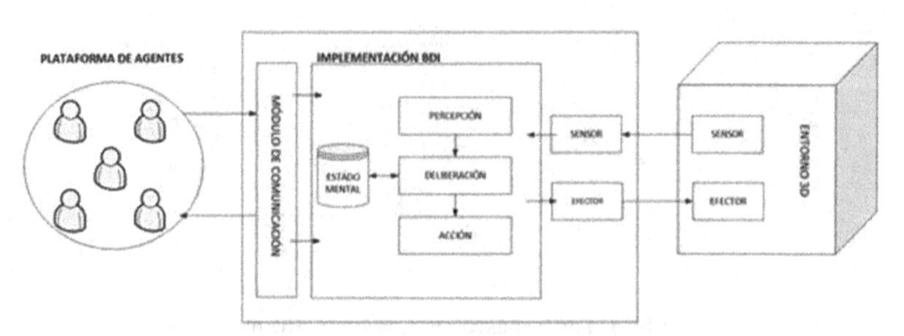

Fig. 2. Agent structure overview

On the other hand, **the 3D environment** consists of an **editor** and a **visualizer**, both implemented using the Unity 3D engine –http://unity3d.com/–. Firstly, the 3D edition environment allows to model virtual tridimensional environments of urban areas (buildings, flats, apartments, gardens, parks, etc.) with a high level of quality and realism. All the maps and 3D models are stored remotely in a *Cloud Computing* (CC) platform. Secondly, the 3D visualizer makes possible the representation and rendering of both the environment and the characteristics that we have defined previously on the 3D editor, which we will get from the CC platform. It will also be responsible for displaying the agents which are part of the system, enabling the visualization of how each agent performs different tasks.

One of the main problems that we can find on the development of 3D scenes can be found when a model needs to move from one point to another one. In such cases, it is necessary to use an algorithm responsible for the path finding. On this occasion we make use of CritterAI [8], a navigation tool for use with Unity. It includes navigation mesh generation, pathfinding and local steering features, including implementations of both A* and Dijsktra search algorithms.

3.2 Communication between the Agents Platform and the Visualization Tool

As we have highlighted previously, one of the main tasks in the platform is to establish communication between JADE and the 3D environment. On this way, we can offer a real corresponding between the events that take place at the platform of agents and the visualization, which occurs at the 3D platform. Inside the platform of agents, there will be an agent who is responsible of managing the communications from JADE.

With this purpose, communication between the platform of agents and the 3D environment will be done through TCP sockets, so with that purpose, it has been developed an Unity's module that manages the requests from the platform of agents and vice versa. On the side of JADE, an agent called Communication agent will be responsible of attending all the requests that will arrive from Unity and doing then the related tasks, attending to the exchanged frame.

Creation, elimination and interaction tasks between agents that are executed at the platform, must be updated at the 3D environment, and the creation, elimination and interaction tasks between agents that are requested from the 3D interface, must be carried out firstly at the platform of agents to be subsequently updated at Unity. The communication will be divided in two blocks:

- Sockets to simulate the JADE tasks and its corresponding update at Unity, with this purpose three elements are used: one socket to send the tasks, another one for confirming the tasks which have been carried out and one table for keeping a record.
- Sockets in charge of performing the tasks which are initialized from the Unity interface, which must be previously done in JADE to be later updated in Unity.

4 Case of Study

The proposed simulation platform has been evaluated through a case of study, which is oriented to validate the proposed system. The scope of this case of study is the analysis of the accessibility level of different workstations in a real company, called Indra Software Labs, located in Salamanca, Spain. Through the performance of several simulations, we will be able to extract knowledge about how disabled workers can find different accessibility problems in the company during their usual activities.

In order to perform the simulations, it will be necessary to define the characteristics of the VO of agents, allowing to model de different processes which take place in the company as faithful to reality as possible. To that end, we will define all the different roles which agents can acquire, the necessary services for the correct performance of the organization, the norms which will govern the society and the messages and interactions between agents. In addition, we have to model the environment where a simulation takes place, in this case the building of the company under consideration. Subsequently, we will carry out a set of simulations in order to show how the organization behaves in different conditions, to finally effect a validation of the proposed model.

First, we define the interaction model, analyzing the needs of the system users, as well as the way in which the information exchange is carried out. De roles we define are as follows:

- **Direction:** this agent is responsible of planning the tasks which the whole organization will carry out. It generates the tasks which represent the objectives to be accomplished by the organization, delegating its distribution to the different agents in charge of the company areas, according to the nature of the task.
- **Area/department responsible:** an agent in charge of planning the distribution of the tasks which have been assigned to the department which it manages, attending to the availability and capabilities of the agents.
- **Worker:** agents that represent each one of the workers participating during the simulation. This kind of agent will have a stack where to store all the tasks that they have to carry out, and have been previously assigned through the area/department responsible. This agents will emit information about the degree of success which they have done each one of the tasks. All this agents share certain elements, which define a set of shared behaviors and cognitive capabilities between them. Beside of this shared characteristics, the agents will have their own characteristics, which are defined according its role in the organization, as well as its disabilities. The assigned roles will determine the tasks which an agent could do, as well as the specific behaviors for each one of them, enabling the possibility of modeling different executions of the same task, attending to the kind of disability that a worker might present.
- **Environmental agents:** represent those elements of the building which an agent can interact with in a direct way. We can find examples of environmental agents in elements such as telephones, lifters or photocopiers.

Among the processes which take place in the company that we are studying, we have identified three kinds of departments, according to the tasks that workers perform: reception, administration and maintenance. The worker-kind agents will have associated a set of tasks that they can carry out, attending to the department which they belong to. Once the building model and the interaction model are defined, we will be able to perform the simulation. To that end, we shall deploy the visualization tool, which will need to perform certain operations before starting the simulation: (i) loading the building model, (ii) generating the navigation mesh, (iii) synchronization with the platform of agents. In Fig. 3 it is shown a visualization tool screenshot.

Fig. 3. 3D representation of the Social simulation

First there has been planned a 15 minutes simulation, with a total of 8 worker agents: 1 receptionist, 2 maintenance and 5 administration workers. Three of them present reduced mobility problems, three of them are visually impaired and one of them is deaf.

After the simulation, the tool generates a report which details all the accessibility problems which avatars have found, and have hampered the correct consecution of the tasks which were assigned to them. 118 tasks have been carried out correctly and 79 have failed (Fig. 4 (A)). Some of the limitations found by the system were: (i) lack of adapted bathrooms for disabled people in common areas, (ii) the workstations are not adapted to deaf or visually impaired people, making impossible to carry out tasks such as answering the phone, or being aware that the fire alarm is active, (iii) lack of access ramps to higher floors as an alternative to the lifter, which, in case of a breakdown will impede the access and exit to the building, (iv) limited access to certain workstations –spaces are not wide enough for a wheelchair– or (v) not accessible furniture elements –filing cabinet, photocopiers-.

Once this accessibility problems have been detected, a redistribution of the furniture elements whose accessibility may be improved, as well as the relocation of certain workstations of people using wheelchairs, or the adapting of certain workstations of people with visual or hearing impairment is proposed.

After this steps, we come to execute a new 15 minutes simulation, obtaining as result a total of 163 successful tasks and 28 failed tasks (Fig. 4 (B)), passing from a reduction of the 40,1% of the failed tasks in the initial simulation, to a 14,7% in the second simulation. Obviously, there are still some elements of the environment which will keep limiting its own accessibility, as the lack of access ramps or adapted bathrooms, but it has been proved that there is a possibility to improve the accessibility of the environment through the restructuration of the furniture or the workstations in the critical points which have been detected through the simulation.

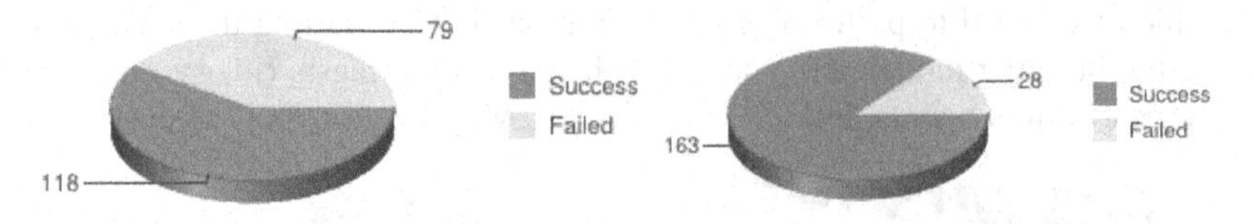

Fig. 4. A: First simulation results; B: Second simulation results

5 Conclusions

Once the simulation platform has been presented and evaluated, we can affirm that it allows performing simulations that are oriented to the integration of disabled people into the workplace, allowing the analysis of both the tasks which are done in the workplace and the environment physical characteristics where this tasks are performed. In this regard, a previous analysis about the state of art on related systems and models, as well as current legislation on this issue has been done.

Moreover, we have model a real organization through the use of a VO of agents, with self-adaptive capabilities. This system has been implemented using JADE, but adapting this platform to our special needs. In the same way, we have created a 3D environment, which is able to model and carry out simulations in the working environment. Additionally, we have developed a system which allows the communication between both platforms, enabling to share relevant information regarding the communications. In conclusion, this implementation as a whole allows simulating the processes which take place in a company, with the purpose of establishing which accessibility problems can occur among the working environment where the represented workers perform their work, as it has been shown on the case of study.

Acknowledgements. This work is supported by the Spanish government (MICINN) and European FEDER funds, project iHAS: Intelligent Social Computing for Human-Agent Societies (TIN2012-36586-C03-03).

References

1. Bellifemine, F., Poggi, A., Rimassa, G.: JADE–A FIPA-compliant agent framework. Proceedings of PAAM 99(97-108), 33 (1999)
2. Borshchev, A., Filippov, A.: From system dynamics and discrete event to practical agent based modeling: reasons, techniques, tools. In: Proceedings of the 22nd International Conference of the System Dynamics Society, vol. (22) (2004)

3. Zato, C., Villarrubia, G., Sánchez, A., Barri, I., Rubión, E., Fernández, A., Rebate, C., Cabo, J.A., Álamos, T., Sanz, J., Seco, J., Bajo, J., Corchado, J.M.: PANGEA – platform for automatic coNstruction of orGanizations of intElligent agents. In: Omatu, S., Paz Santana, J.F., González, S.R., Molina, J.M., Bernardos, A.M., Rodríguez, J.M.C. (eds.) Distributed Computing and Artificial Intelligence. AISC, vol. 151, pp. 229–240. Springer, Heidelberg (2012)
4. Charter of Fundamental Rights of the European Union (2010/C 83/02)
5. Chung, C.A. (ed.): Simulation modeling handbook: a practical approach. CRC Press (2003)
6. Communication from the Commission to the European Parliament, the Council, the European Economic and Social Committee and the Committee of the Regions: European Disability Strategy 2010-2020: A Renewed Commitment to a Barrier-Free Europe (COM/2010/0636)
7. Consolidated version of the Treaty on European Union and the Treaty on the Functioning of the European Union 2012/C 326/01
8. CritterAi Documentation page, `http://www.critterai.org/projects/cainav/doc/` (last access: January 27, 2015)
9. Mulfari, D., Celesti, A., Villari, M.: A computer system architecture providing a user-friendly man machine interface for accessing assistive technology in cloud computing. Journal of Systems and Software 100, 129–138 (2015) ISSN 0164-1212
10. Davidsson, P.: Agent based social simulation: A computer science view. Journal of Artificial Societies and Social Simulation 5(1) (2002)
11. Davidsson, P.: Multi agent based simulation: Beyond social simulation. In: Moss, S., Davidsson, P. (eds.) MABS 2000. LNCS (LNAI), vol. 1979, pp. 97–107. Springer, Heidelberg (2001)
12. Dignum, V.: Handbook of research on multi-agent systems: semantics and dynamics of organizational models, p. 111. Information Science Reference, Hershey (2009)
13. García, E., Rodríguez, S., Martín, B., Zato, C., Pérez, B.: MISIA: Middleware infrastructure to simulate intelligent agents. In: Abraham, A., Corchado, J.M., González, S.R., De Paz Santana, J.F. (eds.) International Symposium on Distributed Computing and Artificial Intelligence. AISC, vol. 91, pp. 107–116. Springer, Heidelberg (2011)
14. European Council Directive 2000/78/EC of 27 November 2000
15. de la Prieta, F., Pérez-Lancho, B., Francisco De Paz, J., Bajo, J., Corchado, J.M.: Ovamah: Multiagent-based adaptive virtual organizations. In: 12th International Conference on Information Fusion, FUSION 2009, pp. 990–997 (2009)
16. Prenkert, F., Følgesvold, A.: Relationship strength and network form: An agent-based simulation of interaction in a business network. Australasian Marketing Journal (AMJ) 22(1), 15–27 (2014) ISSN 1441-3582
17. International Labour Organization, Code of practice on managing disability in the workplace. Document TMEMDW/2001/2 (2001)
18. Bajo, J., Fraile, J.A., Pérez-Lancho, B., Corchado, J.M.: The THOMAS architecture in Home Care scenarios: A case study. Expert Systems with Applications 37(5), 3986–3999 (2010)
19. Corchado, J.M., Corchado, E.S., Aiken, J., Fyfe, C., Fer-nandez, F., Gonzalez, M.: Maximum likelihood Hebbian learning based retrieval method for CBR systems. In: Ashley, K.D., Bridge, D.G. (eds.) ICCBR 2003. LNCS, vol. 2689, pp. 107–121. Springer, Heidelberg (2003)
20. Masmoudi, M.S., Klabi, I., Masmoudi, M.: Fuzzy logic control and HMI interfaces on an intelligent wheelchair system. In: Proceedings-Copyright IPCO, pp. 253–260 (2014)

21. Schreinemachers, P., Berger, T.: An agent-based simula-tion model of human–environment interactions in agricultural systems. Environmental Modelling & Software 26(7), 845–859 (2011) ISSN 1364-8152
22. Rao, A.S., Georgeff, M.P.: BDI Agents: From Theory to Practice. In: ICMAS, vol. 95, pp. 312–319 (June 1995)
23. Almeida, S.J.D., Ferreira, R.P.M., Eiras, Á.E., Ober-mayr, R.P., Geier, M.: Multi-agent modeling and simulation of an Aedes aegypti mosquito population. Environmental Modelling & Software 25(12) (December 2010)
24. Rodriguez, S., Julián, V., Bajo, J., Carrascosa, C., Botti, V., Corchado, J.M.: Agent-based virtual organization architecture. Engineering Applications of Artificial Intelligence 24(5), 895–910 (2011)
25. United Nations. Standard Rules on the Equalization of Opportunities for Persons with Disabilities (1994)
26. United Nations: Factsheet on persons with disabilities, `http://www.un.org/disabilities/default.asp?id=18` (last access: January 27, 2015)
27. Yu, Y., Kamel, A.E., Gong, G., Li, F.: Multi-agent based modeling and simulation of microscopic traffic in virtual reality system. Simulation Modelling Practice and Theory 45, 62–79 (2014) ISSN 1569-190X

Integration of a Purchase Porter System into a VO Developed by PANGEA

Katsumi Harashima[1], Hiroki Nakazaki[1], Pablo Chamoso[2], Javier Bajo[3], and Juan M. Corchado[2]

[1] Osaka Institute of Technology, Osaka, Japan
[2] Computer and Automation Department, University of Salamanca, Spain
[3] Department of Artificial Intelligence, Technical University of Madrid, Spain
{harashima,e1611071}@oit.ac.jp,
{chamoso,corchado}@usal.es, jbajo@fi.upm.es

Abstract. Practical applications of multi-agent systems have acquired a growing relevance during last years. This kind of system is based on distributed artificial intelligence and they pay special attention to organizational concepts, This allows a more realistic design of artificial societies. In this paper, we are presenting a multi-agent system based on virtual organizations, which has been specially designed to model a Purchase Porter System aimed at helping clients of shops to transport their shopping. The system establish agreements to transport shopping to the more appropriated exists for the client. The system has been tested in a simulated environment and the initial results are promising.

Keywords: purchase porter system, multi-agent systems, virtual organizations.

1 Introduction

Even though e-commerce is the most popular commercial method over the last few years, most consumers still go shopping at commercial centres.

A large part of the success of these shopping centres is based on the wellbeing offered to clients. According to [1], six are the main factors that affect comfort: functionality, convenience, safety, leisure, atmospherics and self-identification.

However big malls have disadvantages. For example, clients usually report they find difficult taking bags and shopping items from one shop to another. Obtaining a solution for this problem requires designing new mechanisms for the acquisition and management of huge amounts of information about the environment.

Multi agent systems (MAS), used as a way to create artificial societies, have achieved great importance during the last years [2][3].

MAS are one kind of distributed artificial intelligence in which autonomous and individual agents collaborate and cooperate within an artificial society to achieve defined objectives. These MAS incorporate advanced reasoning models as well as learning and adaptation capacities.

One of their many characteristics is the ability to represent interaction models, close to the ones found in human societies [4]. Nonetheless, they do not consider organizational aspects such as rules or regulations.

J. Bajo et al. (eds.), *Trends in Prac. Appl. of Agents, Multi-Agent Sys. and Sustainability*,
Advances in Intelligent Systems and Computing 372, DOI: 10.1007/978-3-319-19629-9_8

Therefore, it is necessary to utilize virtual agent organizations to implement a realistic artificial society. These organizations allow to represent regulatory and management aspects, which is a much better implementation, being a more efficient and practical solution.

In the framework of this paper, it is proposed a help system to aid customers in shopping malls, designed as a MAS based on virtual organizations (VO), in which every virtual organization represents each different area of the commercial centre. At least, one porter agent will exist in each organization and it will be responsible for receiving all customer purchases, as well as delivering them in a convenient place at a scheduled time.

To accomplish the implementation of this system, PANGEA [5], an agent platform, has been chosen as it was developed by the participants of this work. This is important, as it was mentioned before, for resembling the system architecture to the real structure of shopping malls. Although, virtual organizations are self-adapting systems that suit specific situations automatically, without needing human interaction.

Through this article, the problem described in section 2 will be explained with more detail. In chapter 3, the suggested solution, based on virtual organizations, is introduced and then, its results and conclusions in section 4.

2 Review of the State of the Art

We refer to MAS when two or more agents are able to collaborate for solving a problem [6]. Concretely, one of them is required to be autonomous, and at least, there must exist a relation between two agents in which one of those satisfies the objectives of the other.

The main characteristic is that every agent focus in its own individual task, although the data they use is decentralized across the system. So, each agent must establish their goals and plans to meet their objectives, always taking into consideration the environment where it is located. However, to accomplish those tasks, agents have to work coordinately, so this is why the global system needs to implement communication and control mechanisms.

In order to implement global system control, modelling its structure as a set of VOs can be very helpful. VO are sets of individuals and institutions that need to coordinate their resources and services within some institutional bounds [7]. Therefore, agent-based virtual organizations hold a social structure and a set of rules which regulate the interaction among the different agents. In addition, they have to own a coordination mechanism to determine the way agents fulfill their tasks.

There are several previous works which focus their case studies in the application of models with agent-based virtual organizations to shopping malls, i.e. the one explained in reference [8]. In that case, a MAS was responsible of managing and planning routes to clients making use of THOMAS architecture [9].

3 A VO-Based Purchase Porter Multi-agent System

In the last few years, many large-scale commercial institutions have found important business opportunities by creating shopping malls with a great number of brand stores and services such as cinemas, restaurants or skating rinks. This is one of the reasons why malls are successful.

When people go shopping at commercial centres, they buy products in different stores, carrying more and more items from one shop to another. As purchases increase, customers feel tired and uncomfortable.

For the purpose of solving the problem mentioned above, we have proposed a purchase porter system with multi-agent technology. In this porter system, individual porter agents have their own workspace. They receive purchases from each store and then, they carry them to the exit to be on time for the desire collection time.

When a porter agent of a certain area receives a transportation request from a shop, products are delivered from the shop location to the agent position. If clients desire to pick the items at the exit of this area, the porter agent transport them to the specified place on time.

Otherwise, the agent communicates with the representatives of adjacent areas. Internally, the best possible route is calculated considering the destination, time remaining and the priority of other requests.

This way, products are optimally carried across several areas by delegating the transportation task among several porter agents, each one responsible of its own workspace.

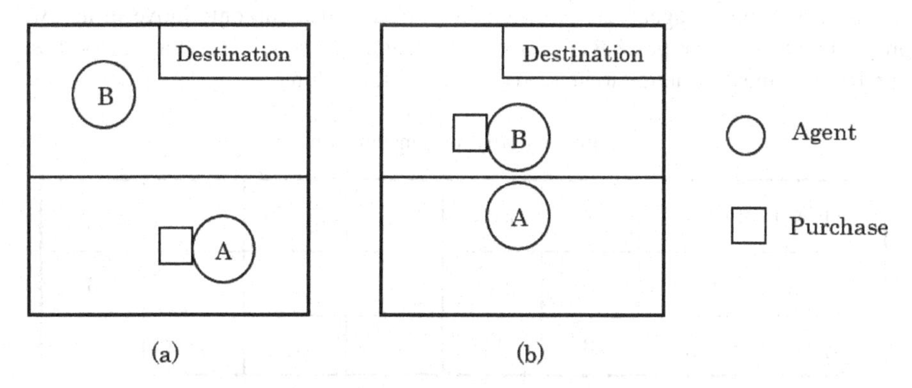

Fig. 1. Transfer of purchase

Figure 1 shows the transfer of purchases among agents. In Fig 1(a), the initial situation is represented, having the agent A, a purchase needed to be delivered at the destination point. As this location is at the exit of agent B workspace, A gives the purchase to agent B, Fig 1(b).

With this system, clients do not need to take items with them every time they enter a new shop. Also, purchases are collected whenever the client wants and in the exit they choose. Usually, customers would like to pick all the items when they return home.

3.1 Features of the Purposed Porter System Are as Follows:

3.1.1 Environment

The environment has several stores and exits.

- Store

  In each store, transportation of purchases occurs constantly.

- Destination

  Every destination is located on the outer circumference of the environment and they become the final destination of the purchase.

3.1.2 Purchase

Every purchase has information about a collection point (exit) and time, both set by the customer. Our system adds a priority calculated according to the remaining time and distance to the destination.

Table 1 shows priority for the destination time. Our system attaches a priority so the system is working full time. The shorter the remaining time, the higher the priority is.

Table 2 shows priority for the distance to the destination. It includes a priority system so that the distance becomes as long as possible. When a distance is shorter, the associated priority to the distance is larger, due to efficiency.

Porter agents are able to transport purchases by evaluating these priorities without being late for a time limit.

3.1.3 Agent

An individual porter agent has a certain workspace and it can only move in its own area. They receive the purchases of a shopper from each store. Then they carry those purchases to the destination to be ready for collection on time.

Table 1. Priority for remaining time

remaining time t	$t > 50$	$50 > t$	$40 > t$	$30 > t$	$20 > t$	$10 > t$
m	1	2	3	4	5	6
time limit $f(m)$	3	11	26	50	85	133

Table 2. Priority for distance

distance l	$l > 10$	$10 > l$	$8 > l$	$6 > l$	$4 > l$	$2 > l$
n	1	2	3	4	5	6
distance priority $g(n)$	2	8	20	40	70	112

Each agent behaves as described below:

1. First, it receives information about a new purchase (transport request) from a shop in its workspace.
2. It compares the new purchase priority with the others queued.
3. If the new priority is higher, it has to move to the shop which sent the information and receive the items. Otherwise, it does not do anything.
4. If the destination of the purchase with the highest priority is within the agent's workspace, then it takes it there. Otherwise, jump to step 6.
5. If a new purchase in its workspace is received, jump to step 1.
6. If the adjacent agents are located near the destination, it gives the purchases to those agents only if the requests in their queues have lower priorities.
7. If an agent does not have any purchase, it moves to the center of its workspace and then starts again from step 1. If not, it jumps to step 5.

3.2 Integration in a Multi-agent Platform (PANGEA)

To implement this system, it has been decided to use multiagent architecture PANGEA as it is based in organizational concepts. This, regarding its similarity with the real world, is going to permit modelling and implementing the large range of different structures shopping malls have. Although, as a rule, they usually share similar organizations.

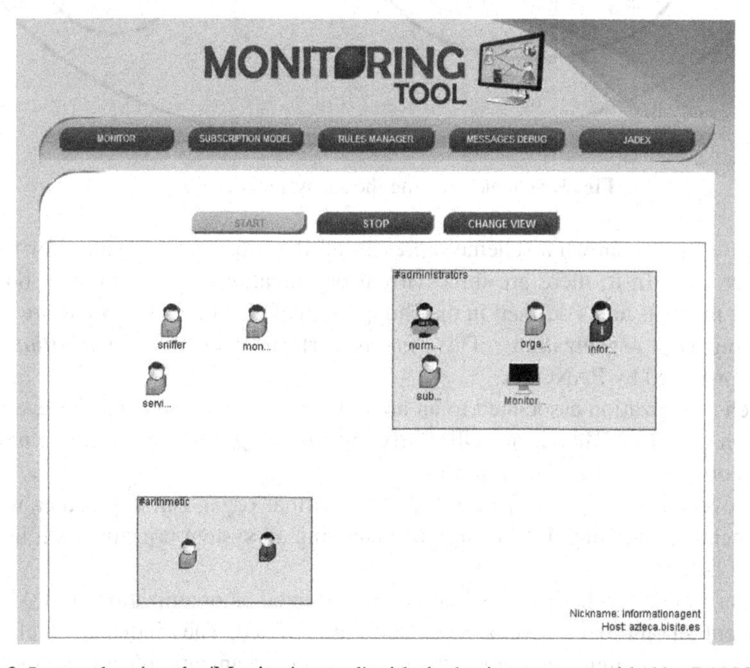

Fig. 2. Image showing the 'Monitoring tool' with the basic agents provided by PANGEA

This characteristic allows to improve the control of each node in the system in an individual or collective way.

Furthermore, this is fundamental when controlling and monitoring the state of the client purchases in every moment. This process is accomplished by the 'Monitoring Tool' which allows to keep and visualize records of everything that has happened in the system. It also enables to define a set of rules which are fundamental during the process of implementation.

Several specifically designed and structured agents have been added to control agents, which the architecture presents for using basic functions. These new agents develop the case study introduced in this article.

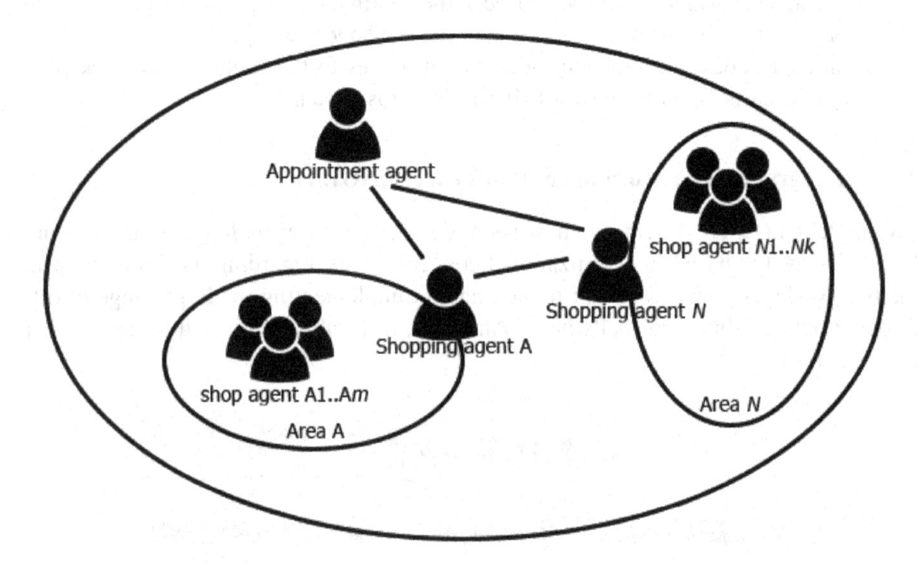

Fig. 3. Scheme showing the subsystem developed

In figure 3, it is shown a scheme representing the subsystem structure for the case study proposed. In it, there are three virtual organizations as an example, but there will be as many as areas defined in the shopping mall plus one. New areas are defined by creating new organizations. This task is performed by the ***organization agent*** which is provided by PANGEA.

In each organization associated to an area, there must exist an agent for every shop existing in that area. This agent will notify the ***shopping agent*** each time a new purchase is completed in the shop it represents.

Also, every shopping agent is located in a virtual organization together with the appointment agent, which is in charge of managing all system priorities as it has been explained before.

In order to check whether which areas (represented as organizations in PANGEA) are adjacent among them, several XML rules are defined. They indicate which organizations can communicate among them (adjacent organizations) and those which cannot. This process is carried out by the PANGEA rule agent, using its monitoring tool following the scheme below:

```
<norm id=0 type="prohibition">
        <organization name="*"/>
        <organization name="*"/>
        <level>10</level>
</norm>
<norm id=1 type="permission">
        <organization name="A"/>
        <organization name="B"/>
        <level>1</level>
</norm>
<norm id=2 type="permission">
        <organization name="A"/>
        <organization name="C"/>
        <level>1</level>
</norm>
...
```

This would mean, as expressed in the rule with identifier equal to 0, that no organization could communicate with others, unless there would be rules with higher priority (lower level) indicating the opposite. This is the case for rules with id 1 and 2 which express that organizations (areas) A-B and A-C are adjacent.

4 Results and Conclusions

This article introduces a new porter system with multi-agent systems and virtual organizations. The main focus of this article is to provide a customer-driven solution in large scale commercial institutions. However, satisfying all clients may be a complicated task, due to its high number. Despite this, the system is able to transport almost all purchases to the exit on the specified time. This is possible thanks to the proposed priority queue.

However, in our system, the stores request delivery of purchases at a constant interval. But this is unrealistic. Therefore, our future work is to implement several improvements to address a real flow of purchases over time.

Regarding the use of virtual organizations, the architecture is organized in a way that resembles a real shopping mall. This way, areas and agents are distributed easily. Likewise, PANGEA features fit perfectly to create organization-based structures, permitting control and monitoring all moves occurred. This architecture presents an adaptive and scalable design capable of adjusting itself to new conditions, being applicable to any commercial setting.

References

[1] El Hedhli, K., Chebat, J.C., Sirgy, M.J.: Shopping well-being at the mall: Construct, antecedents, and consequences. Journal of Business Research 66(7), 856–863 (2013)
[2] Lane, J.E.: Method, Theory, and Multi-Agent Artificial Intelligence: Creating computer models of complex social interaction. Journal for the Cognitive Science of Religion 1(2), 161–180 (2014)
[3] Fan, Z., Duan, W., Chen, B., Ge, Y., Qiu, X.: Study on the method of multi-agent generation algorithm within special artificial society scene. In: 2012 UKACC International Conference on Control (CONTROL), pp. 1076–1081. IEEE (September 2012)
[4] Schreinemachers, P., Berger, T.: An agent-based simulation model of human–environment interactions in agricultural systems. Environmental Modelling & Software 26(7), 845–859 (2011)
[5] Zato, C., Villarrubia, G., Sánchez, A., Bajo, J., Corchado, J.M.: PANGEA: A New Platform for Developing Virtual Organizations of Agents. International Journal of Artificial Intelligence™ 11(A13), 93–102 (2013)
[6] Mas, A.: Agentes software y sistemas multiagente: conceptos, arquitecturas y aplicaciones. Prentice Hall (2005)
[7] Boella, G., Hulstijn, J., Van Der Torre, L.: Virtual organizations as normative multiagent systems. In: Proceedings of the 38th Annual Hawaii International Conference on System Sciences, HICSS 2005, pp. 192c–192c. IEEE (January 2005)
[8] Rodriguez, S., Julián, V., Bajo, J., Carrascosa, C., Botti, V., Corchado, J.M.: Agent-based virtual organization architecture. Engineering Applications of Artificial Intelligence 24(5), 895–910 (2011)
[9] Julian, V., Rebollo, M., Argente, E., Botti, V., Carrascosa, C., Giret, A.: Using THOMAS for Service Oriented Open MAS. In: Kowalczyk, R., Vo, Q.B., Maamar, Z., Huhns, M. (eds.) SOCASE 2009. LNCS, vol. 5907, pp. 56–70. Springer, Heidelberg (2009)

Part III

Special Sessions on Multi-Agent Systems and Ambient Intelligence (MASMAI)

Reasoning about Multi-Agent Systems Using Stochastic Petri Nets*

Bruno Lopes[1], Mario Benevides[3], and Edward Hermann Haeusler[2]

[1] Departamento de Ciência da Computação, Universidade Federal Fluminense, Niteri-RJ Brazil
blopes@pq.cnpq.br,
http://www.tecmf.inf.puc-rio.br/BrunoLopes

[2] Departamento de Informática, Pontifícia Universidade Católica do Rio de Janeiro, Rio de Janeiro-RJ Brazil
hermann@inf.puc-rio.br
http://www.tecmf.inf.puc-rio.br/EdwardHermann

[3] Programa de Engenharia de Sistemas e Computação
Universidade Federal do Rio de Janeiro, Rio de Janeiro-RJ, Brazil
mario@cos.ufrj.br
http://www.cos.ufrj.br/~mario/

Abstract. Multi-agent systems are composed by many independent agents where the task of some of them may depend on the task of others. In this work we present a generic Stochastic Petri Net model for agents and illustrate the usage of a sound, complete and decidable logic system to verify properties in multi-agent systems modelled as Stochastic Petri Nets: the $\mathcal{DS}_3$ logic. This logic takes advantage of the intuitive graphical interpretation of Petri Nets, allowing the user to model the behaviour of agents and their interactions by means of nets (i.e. seen Petri Nets as graphs). Our approach leads not only to a usual place-transition connection provided by Petri-nets underlying graphs modelling, such as the verification of properties and validation of agents, but also regards the verification of properties concerning their behaviour inside an environment.

1 Introduction

Multi-agent systems are composed by several agents that aim to eventually reach some goals that need the contribution of other agents [1, 2]. The agents are autonomous and there is no centralised controller which leads to a complex scenario to verify properties into the full system. Some of the difficulties lay down on how to deal with the asynchronous aspects of the concurrency.

Into the task of verifying properties it is imperative to consider the interaction between agents which may lead to scenarios of higher complexity. Among many other characteristics, it may be required to verify if some situation may be achieved (reachability), if two or more processes are waiting for each other

* The authors thank CAPES, CNPq and FAPERJ for the partial support of this research.

J. Bajo et al. (eds.), *Trends in Prac. Appl. of Agents, Multi-Agent Sys. and Sustainability*, Advances in Intelligent Systems and Computing 372, DOI: 10.1007/978-3-319-19629-9_9

to release some resource (deadlock), and if the system will be working properly (liveness). This also regards synchronization aspects that are vital to ensure the safety of systems. However, this seems to be still a problem nowadays. The verification procedures may lead to callbacks, deadlocks and state explosions [3, 4]. The interaction among the agents are generally described as depending on time. This allows its definition by means of Stochastic Processes [5].

Stochastic Processes lack on natural ways to model synchronization, which are desired for many multi-agent scenarios. Concerning situations where many agents need to communicate sharing or contending resources, synchronisations are always present and deal with system properties quite difficult to verify.

Petri Nets are a formalism to specify concurrent systems which has an intuitive graphical interpretation. Agents are independent but may require the result of the computations of other agents (collaboratively or competitively) to reach some goal. In a multi-agent system, that can achieve a huge amount of variables, this requirement may lead to a quite difficult scenario to predict. The usage of Petri Nets can offer means to model such systems. Model multi-agent systems using Petri Nets takes a large body of literature [2, 6–9].

In order to increase the expressiveness of Petri Nets, we associate an exponentially distributed random variable to each transition, defining a Stochastic Petri Net (SPN – detailed in Section 2). This model of Petri Nets is widely used for non-linear time modelling and, as ordinary Petri Nets, it has an intuitive graphical interpretation and provides native support to synchronization. We extend the model by associating a random variable with each transition to control their firing rates (i.e. making it possible to model random subjects of the environment) [10–13].

The usage of SPN to model multi-agent systems is already present into the literature [9, 14, 15]. Not only by the increased expressive power, in scenarios where state explosion is present, it may be quite difficult to achieve the optimal answer on the formal verification. Using SPN it is possible to get a probability answer which may give a confidence of safety, which is inherited from its Stochastic Process behaviour.

We present the usage of a logic to deal with Stochastic Petri Nets, the $\mathcal{DS}_3$ logic. The main advantage of this approach is to lead the user the capability of achieving logical certification of properties as well as a probabilistic answer for their verification. This work shows how one can model and verify properties of multi-agent systems using this logic.

The $\mathcal{DS}_3$ logic [16] (detailed in Section 4) is a sound, complete and decidable extension of Petri-PDL [17]: a muti-modal logic used for reasoning about Petri Net programs where each modality is composed of a program π with a markup s, denoting a modality as $\langle s, \pi \rangle$. The interpretation of a formula $\langle s, \pi \rangle \varphi$ is that after the Petri Net program π runs with initial markup s there is a state reachable from the initial state where φ holds (and analogous for a "necessity" modality where φ must hold in all states accessible from the initial: $[s, \pi]\varphi \leftrightarrow \neg\langle s, \pi \rangle \neg\varphi$). The extension presented by $\mathcal{DS}_3$ concerns in replace the ordinary Petri Net program by an Stochastic Petri Net program, increasing its expressability.

Despite other stochastic approaches to Dynamic Logics for reasoning about concurrent systems [18–22], $\mathcal{DS}_3$ presents good features. The other systems found in the available literature lack on some properties as a finite axiomatization, allow boolean combination of propositional variables, a better expressability (some are defined only for regular programs), decidability, may require a translation of the program to a measurable function (hard to model) or only let one to know if some probability is greater than a constant value.

Combining the usage of Stochastic Petri Nets and $\mathcal{DS}_3$ we propose a methodology to formally very properties in multi-agent systems. SPNs naturally encode the concurrency aspects and $\mathcal{DS}_3$ allows the verification of satisfiability of properties, leading this approach to a huge theoretical support for properties verification that are not expressible in other formalisms (i.e. stochastic processes) or not verifiable (i.e. trace theory). This work extends [23] by considering a generic Petri Net model for agents [24] (extending it in order to take a more general model). One of the advantages of our approach is its ability to verify properties not only into the interactions between agents but also verifying properties and validate agents. Our contribution is to show, by means of a concluding conceptual argument, that Stochastic Process and Logic based validation work together better than individually on providing tools for MAS formal validation.

2 Stochastic Petri Nets

A Stochastic Petri Net (SPN) [13] is a tuple $\mathcal{P} = \langle S, T, W, M_0, \Lambda \rangle$, where S is a finite set of places, T is a finite set of transitions with $S \cap T = \emptyset$ and $S \cup T \neq \emptyset$ and W is a function which defines directed edges between places and transitions and assigns a $w \in \mathbb{N}$ that represents a multiplicative weight for the transition, as $W\colon (S \times T) \cup (T \times S) \to \mathbb{N}$ (in this work we consider w always as 1), M_0 is the initial markup and $\Lambda = \lambda_1, \lambda_2, \ldots, \lambda_n$ the firing rates of each transition (i.e. the parameter of each exponentially distributed random variable associated with each transition, a way of model the proportion of firing of this transition despite its preset).

The firing rate in an SPN is determined by the markups and the firing rate. Each transition $t_i \in T$ is associated to an unique random variable of the exponential distribution with parameters $\lambda_i \in \Lambda$.

In the initial markup (M_0) each transition sets its firing delay (i.e. a timing for an enabled transition fires) by an occurrence of the random variable whose transition is associated. Each firing delay is marking-dependent and the $t_i \in T$ firing rate at marking M_j is defined as $\lambda_i(M_j)$ and has average firing delay (i.e. the average value of a random variable occurrences) of $[\lambda_i(M_j)]^{-1}$. After a firing, each previously non-marking-enabled transition gets a new firing delay by sampling its associated random variable. A previously marking-enabled that keeps marking-enabled has its firing delay decreased in a constant speed. When a transition firing delay reaches zero, this transition fires. In this work we consider the subset of Stochastic Petri Nets described below.

So, given a markup M of a Petri Net, we say that a transition t_i is enabled on M_j if and only if there is a token in each place of its preset and its timing is the minimum into the Petri Net, that is $\forall x \in {}^{\bullet}t_i, M_j(x) \geq 1$ and $\lambda_i(M_j) \leq \min(\lambda_1(M_j), \lambda_2(M_j), \ldots, \lambda_n(M_j))$, where ${}^{\bullet}t_i$ denotes the preset of the transition t_i and $t_i^{\bullet}$ denotes its postset. A new markup generated by setting a transition which is enabled is defined in the same way as in an ordinary Marked Petri Net, i.e.

$$M_{i+1}(x) = \begin{cases} M_i(x) - 1, \forall x \in {}^{\bullet}t \setminus t^{\bullet} \\ M_i(x) + 1, \forall x \in t^{\bullet} \setminus {}^{\bullet}t \\ M_i(x), \quad \text{other case} \end{cases} \tag{1}$$

and a new firing delay for a transition t_i after a markup M_j is defined as

$$\lambda_i(M_{j+1}) \begin{cases} = \text{new}_e(\lambda_i) \text{ if } \begin{cases} \begin{cases} \forall x \in {}^{\bullet}t_i, M_j(x) \geq 1 \\ \lambda_i(M_j) \leq \min(\lambda_1(M_j), \ldots, \lambda_n(M_j)) \end{cases} \\ \text{or} \\ \begin{cases} \exists x \in {}^{\bullet}t_i, M_j(x) < 1 \\ \forall x \in {}^{\bullet}t_i, M_{j+1}(x) \geq 1 \end{cases} \end{cases} \\ < \lambda_i(M_j) \quad \text{other case} \end{cases} \tag{2}$$

where $\text{new}_e(\lambda)$ denotes a new occurrence of the random variable exponentially distributed with parameter λ associated to t_i. Notice that in this work we consider that $\forall x \forall y W(x, y) = 1$.

We follow the restriction presented in the work of Almeida and Haeusler [25] in which there are only three types of transition to define all valid Petri Nets due to its compositions. These basic Petri Nets are as in Fig. 1.

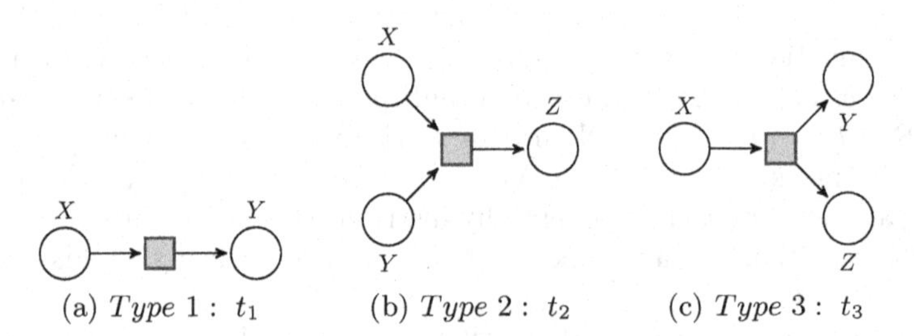

Fig. 1. Basic Petri Nets

3 Multi-agent Systems as Stochastic Petri Nets

To exemplify the modelling and verification of multi-agent systems we consider a scenario composed by agents that will collect some data and process it. This model takes attention to the fact that some agents may be energy limited (e.g. agents inside wireless sensors). Considering the case where energy is not relevant (i.e. the agent is always connected to some power supply), that is the case where there is always a token in the energy representative place.

Take into account the SPN on Fig. 2. The amount of tokens in the place "Energy" denotes the running (working) is possible. A token in other places denotes that the respective process in the stage of processing.

Note that the verification step (i.e. the place "Verify") can not be properly modelled in an Ordinary Petri Net. This step denotes a stage of validation of the correctness of the result (two transitions may fire after this stage, one for success and another for failure). If the verification returns ok then the data is returned by the agent. If not, they send a token to the place "Parse" meaning that the data will begin its processing again. As the failure rate is not (usually) the same than the success rate, it is required to set adequate fire rates (i.e. use an stochastic variable adequate to each transition). The λ parameter associated to the corresponding transition will denote the failure rate of the processing.

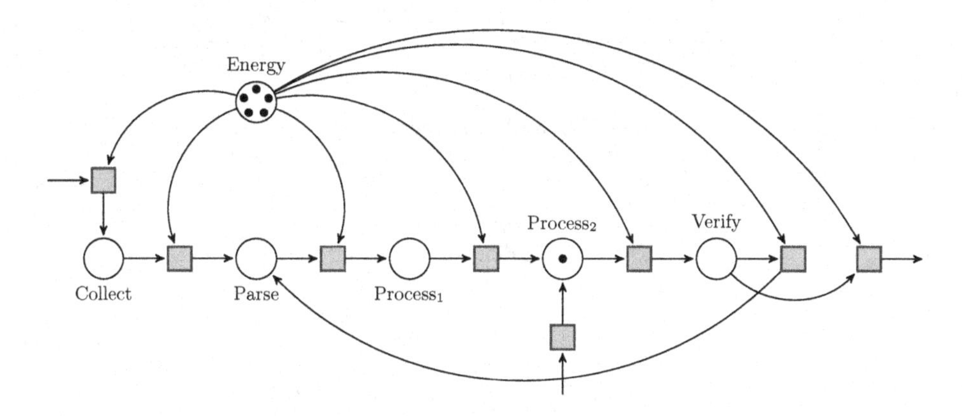

Fig. 2. A complete agent

To compose a hybrid multi-agent system (i.e. with more than one kind of agents) we have another kind of agent, denoted by the SPN in Fig. 3. This SPN denotes a simpler agent in which the data cannot be fully processed. This agent must send its partially processed data to an agent capable of complete its processing (the transition inputing into place "Process$_2$" in Fig. 2 denotes the input of the semi-processed data by this simpler agent). We also consider that these simpler agents have a clock faster than the full agent presented in Fig. 2. To deal with these clock differences we must set adequate values to the λ parameter of the corresponding transitions (i.e. set the firing rate).

This agent model is inspired in a generic Petri Net agent model [24], modifying it in order to represent more details (e.g. the failure rate) and accomplished with a Stochastic Petri Net. This model can be extended to more detailed scenarios decomposing places Process$_1$ and Process$_2$ in SPNs that detail the processes.

We compose these two kinds of agents in the scenario presented in the SPN of Fig. 4. We have four agents (A_1, A_2, A_3 and A_4) that must collect and process some data from the resource centre (r), but agents A_1 and A_2 can not make the full process and need that A_3 or A_4 complete the computation. Another characteristic of this system is that A_3 and A_4 have a faster processor than A_1 and A_2 and that A_1 and A_2 are in a shared memory system, but the clock of the processor of A_1 is faster than A_2. As the clock of the processor of A_1 is faster than the one of A_2, the firing rates (i.e. the λ parameter of the random variable which is associated with the transitions whose preset or postset depends on A_1) is greater than those of A_2.

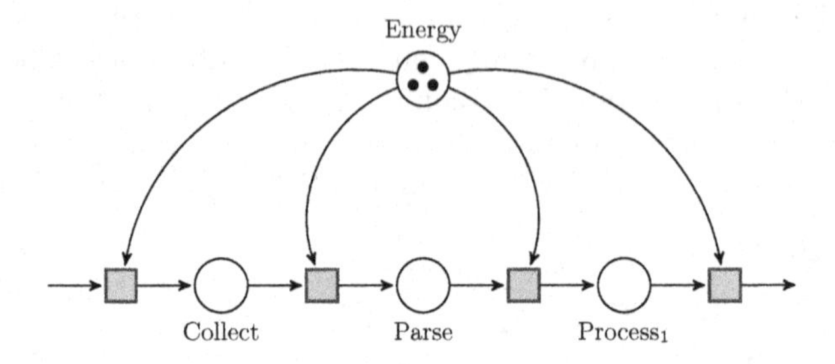

Fig. 3. A basic agent

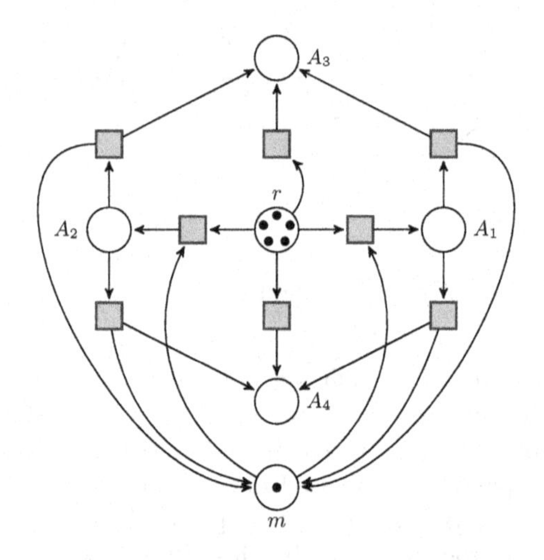

Fig. 4. Petri Net of a four agents system

Fig. 3 denotes the behaviour of agents A_1 and A_2 and Fig. 2 denotes the behaviour of agents A_3 and A_4. Note that to model the fact that the processor of A_1 is faster than the processor of A_2 we may set values to the λ parameter random variables associated to the transitions that represents this fact. Concerning a full agent (i.e. an agent that is able to complete the desirable computation – agents A_3 and A_4), they have a verification step, in which they test if the processing is correct. If it is correct, they just send the data (i.e. the place correspondent in the SPN of Fig. 4 will receive a token).

Notice that the leftmost arrow with no place in its origin denotes the input from the data resource of Fig. 4. Similarly, the bottom most arrow with no place in its origin denotes the input from a basic agent. This connection is used for synchronisation between agents (i.e. a basic agent send its preprocessed data to finalize its processing).

This SPN of Fig. 4 encapsulates the complexity of the internal operations of each agent. Despite of the fact that some properties will not be verifiable, this approach leads to a simple net in which properties verification and/or validation may be efficiently done (e.g. reducing state explosion in a model checking procedure). If it is required to know some property inside some agent, it is possible to "zoom in" into the SPN place, replacing it by the agent SPN.

4 The $\mathcal{DS}_3$ Logic

The logic $\mathcal{DS}_3$ [16] is a multi-modal logic in which each program is a Stochastic Petri Net. Among its advantages, we have that $\mathcal{DS}_3$ is proved to be sound, complete and decidable and has deductive systems.

The $\mathcal{DS}_3$ language consists of

Propositional symbols: p, q. . . , where Φ is the set of all propositional symbols
Place names: e.g.: $a, b, c, d \ldots$
Transition types: $\mathbf{T_1}$:at_1b, $\mathbf{T_2}$:abt_2c and $\mathbf{T_3}$:at_3bc, each transition has a unique type
Petri Net Composition symbol: $\odot$
Sequence of names: $S = \{\epsilon, s_1, s_2, \ldots\}$, where ϵ is the empty sequence. We use the notation $s \prec s'$ to denote that all names occurring in s also occur in s'.

Definition 1. $\mathcal{DS}_3$ program
A $\mathcal{DS}_3$ program is a pair (Π, Λ) where Π is a composition of transitions defined as (let s a sequence of names – the markup of Π):

Basic programs: $\pi_b ::= at_1b \mid at_2bc \mid abt_3c$ *where t_i is of type $T_i, i = 1, 2, 3$*
Stochastic Petri Net Programs: $\pi ::= \pi_b \mid \pi \odot \pi$
Marked Stochastic Petri Net Program: $\Pi = s, \pi$

$\Lambda(\pi) = \langle \lambda_1, \lambda_2, \ldots \lambda_n \rangle$ is a function that associates a positive real value with each basic transition $\pi \in \Pi$ where $\pi = \pi_1 \odot \pi_2 \odot \cdots \odot \pi_n$ that denotes the value of the parameter of the exponential random variable associated with each transition.

Definition 2. *A formula is defined as (let s a sequence of names)*

$$\varphi ::= p \mid \top \mid \neg\varphi \mid \varphi \wedge \varphi \mid \langle s, \pi \rangle \varphi.$$

We use the standard abbreviations $\bot \equiv \neg\top$, $\varphi \vee \phi \equiv \neg(\neg\varphi \wedge \neg\phi)$, $\varphi \to \phi \equiv \neg(\varphi \wedge \neg\phi)$ and $[s, \pi]\varphi \equiv \neg\langle s, \pi \rangle \neg\varphi$.

Definition 3. *We define the firing relation $f : S \times \pi_b \to S$ as follows*

– $f(s, at_1b) = \begin{cases} s_1bs_2, if\ s = s_1as_2 \\ \epsilon, if\ a \not\prec s \end{cases}$

– $f(s, abt_2c) = \begin{cases} s_1cs_2s_3, if\ s = s_1as_2bs_3 \\ \epsilon, if\ a, b \not\prec s \end{cases}$
– $f(s, at_3bc) = \begin{cases} s_1s_2bc, if\ s = s_1as_2 \\ \epsilon, if\ a \not\prec s \end{cases}$
– $f(\epsilon, \eta) = \epsilon$, *for all petri nets programs* η.

Definition 4. *$\mathcal{DS}_3$ Frame*
A frame for $\mathcal{DS}_3$ is a tupple $\mathcal{F}_3 = \langle W, R_\pi, M, (\Pi, \Lambda), \delta\rangle$ where

– *W is a non-empty set of states*
– $M\colon W \to S$
– *(Π, Λ) is a Stochastic Petri Net where*
  - *Π is a finite Stochastic Petri Net such that for any program π used in a modality, $\pi \in \Pi$ (i.e. π is a subnet of Π)*
  - *$\Lambda(\pi) = \langle\lambda_1, \lambda_2, \ldots, \lambda_n\rangle$ is the sequence of $\mathbb{R}^+$ values denoting the fire rate of each transition of $\pi_1 \odot \pi_2 \odot \cdots \odot \pi_n = \pi \in \Pi$*
– *$\delta(w, \pi) = \langle d_1, d_2, \ldots, d_n\rangle$ is the sequence of firing delays of the program $\pi \in \Pi$ in the world $w \in W$ respectively for each program $\pi_1 \odot \pi_2 \odot \cdots \odot \pi_n = \pi$, satisfying the following conditions (let $s = M(w)$ and $r = M(v)$)*
  - *if $wR_{\pi_b}v$, $f(r, \pi_b) = \epsilon$, $\delta(w, \pi_b) = \delta(v, \pi_b)$*
  - *if $f(s, \pi_b) = \epsilon$, $f(r, \pi_b) \neq \epsilon$ and $wR_{\pi_b}v$, $\delta(v, \pi_b)$ is an occurrence of a random variable of exponential distribution with parameter $\Lambda(\pi_b)$*
  - *if $f(s, \pi_b) \neq \epsilon$, $f(r, \pi_b) \neq \epsilon$ and $wR_{\pi_b}v$, $\delta(v, \pi_b) < \delta(w, \pi_b)$*
– *R_α is a binary relation over W, for each basic program $\alpha \in \pi_b$, satisfying the following conditions (let $s = M(w)$)*
  - *if $f(s, \alpha) \neq \epsilon$ and $\delta(w, \alpha) = \min(\delta(w, \Pi))$, $wR_\alpha v$ iff $f(s, \alpha) \prec M(v)$*
  - *if $f(s, \alpha) = \epsilon$ or $\delta(w, \alpha) \neq \min(\delta(w, \Pi))$, $wR_\alpha v$ iff $w = v$*
– *we inductively define a binary relation R_η, for each Petri Net program $\eta = \eta_1 \odot \eta_2 \odot \cdots \odot \eta_n$, as $R_\eta = \{(w, v) \mid \exists\eta_i, \exists u$ such that $s_i \prec M(u)$ and $wR_{\eta_i}u$ and $\delta(w, \eta_i) = \min(\delta(w, \Pi))$ and $uR_\eta v\}$ where $s_i = f(s, \eta_i)$, for all $1 \leq i \leq n$.*

Definition 5. *$\mathcal{DS}_3$ Model*
A model for $\mathcal{DS}_3$ is a pair $\mathcal{M} = \langle\mathcal{F}_3, \mathbf{V}\rangle$, where $\mathcal{F}_3$ is an $\mathcal{DS}_3$ frame and $\mathbf{V}$ is a valuation function $\mathbf{V}\colon \Phi \to 2^W$.

Lemma 1. *Truth Probability of a Modality*
The probability of $\mathcal{M}_3,w \Vdash \langle s, \pi_b\rangle\varphi$ is (let $s = M(w)$)

$$\Pr(\mathcal{M}_3,w \Vdash \langle s, \pi_b\rangle\varphi \mid \delta(w, \Pi)) = \frac{\delta(w, \pi_b)}{\sum\limits_{\pi_b \in \Pi : f(s, \pi_b) \neq \epsilon} \delta(w, \pi_b)}$$

Definition 6. *Semantic notion of $\mathcal{DS}_3$*
Let $\mathcal{M}_3$ be a model for $\mathcal{DS}_3$. The notion of satisfaction of a formula φ in $\mathcal{M}_3$ at a state w, says $\mathcal{M}_3,w \Vdash \varphi$ is inductively defined as follows.

– *$\mathcal{M}_3,w \Vdash p$ iff $w \in \mathbf{V}(p)$;*

- $\mathcal{M}_3,w \Vdash \top$;
- $\mathcal{M}_3,w \Vdash \neg\varphi$ *iff* $\mathcal{M}_3,w \nVdash \varphi$;
- $\mathcal{M}_3,w \Vdash \varphi_1 \wedge \varphi_2$ *iff* $\mathcal{M}_3,w \Vdash \varphi_1$ *and* $\mathcal{M}_3,w \Vdash \varphi_2$;
- $\mathcal{M}_3,w \Vdash \langle s,\eta\rangle\varphi$ *if there exists* $v \in W$, $wR_\eta v$ *and* $\Pr(\mathcal{M}_3,v \Vdash \langle s,\eta_b\rangle\varphi \mid \delta(v,\Pi)) > 0$ *(note that* $\Pi \in (\Pi,\Lambda)$ *is the SPN of the model).*

So if we say that $\mathcal{M}_3,w \Vdash \langle s,\eta\rangle\varphi$ *then it means that the program* η *beginning with the markup* s *has probability of running greater than one (i.e. the probability of a firing happens is greater than zero) and that when it stops* φ *holds in the current state. If* φ *is valid in all states of* $\mathcal{M}_3$ *then* φ *is valid in* $\mathcal{M}_3$*, says* $\mathcal{M}_3 \Vdash \varphi$*; and if* φ *is valid in any model then* φ *is valid, says* $\Vdash \varphi$.

5 Verifying Properties

We come back to the SPN in Fig. 4. Taking a propositional formula p that means that all data was processed, the formula $\langle\{rrrrrm\}, rmt_2A_1 \odot rmt_2A_2 \odot rt_1A_3 \odot rt_1A_4 \odot A_1t_3A_3m \odot A_2t_3A_3m \odot A_1t_3A_4m \odot A_2t_3A_4m\rangle p$ says that after some running of the Petri Net of Fig. 4, p holds, that is, all the data are processed.

Verify if this formula holds in a state w of a model $\mathcal{M}$ (i.e. verify if it is possible that some transition fires) is equivalent to compute the probability of some basic program fire is greater than zero, which is reduced to the equation in lemma 1. To verify if it is possible that A_1 and A_2 compute some data in parallel, we verify that after some of them begin to process something (i.e. rmt_2A_1 or rmt_2A_2 fires), m will not be anymore in the sequence of names, so it is not possible that the other agent starts to compute something unless a transition that restates a token to m fires.

If it is desirable to know if, from a state w, it is possible that some agent (e.g. agent A_1) collect some data to process, it is just needed to compute $\Pr(\mathcal{M},w \Vdash \langle s, rmt_2A_1\rangle\top \mid \delta(w, rmt_2A_1 \odot rmt_2A_2 \odot rt_1A_3 \odot rt_1A_4 \odot A_1t_3A_3m \odot A_2t_3A_3m \odot A_1t_3A_4m \odot A_4t_3A_4m))$, where $s = M(w)$, and verify if it is greater than zero. By lemma 1 it is equivalent to verify if

$$\frac{\delta(w, rmt_2A_1)}{\sum\limits_{\pi_b\in\Pi:f(s,\pi_b)\neq\epsilon}\delta(w,\pi_b)}$$

is greater than zero, where $\Pi = rmt_2A_1 \odot rmt_2A_2 \odot rt_1A_3 \odot rt_1A_4 \odot A_1t_3A_3m \odot A_2t_3A_3m \odot A_1t_3A_4m \odot A_4t_3A_4m$.

A more sophisticated example concerns in verifying if the transmission ratings from agents A_1 and A_2 to the agents A_3 and A_4 are overheading agents A_3 and A_4. That is verify if the programs $A_1t_1A_3$, $A_1t_1A_4$, $A_2t_1A_3$ and $A_2t_1A_4$ are firing more times than rt_1A_3 and rt_1A_4. It is equivalent to verify if the probabilities of firing that first basic programs are greater than these last ones. So it is equivalent to verify if for a sequence $\Lambda(A_1t_1A_3 \odot A_1t_1A_4 \odot A_2t_1A_3 \odot A_2t_1A_4)$ from an initial state v_1 such that $v_1R_\Pi v_n = v_1Rv_2 \circ \cdots \circ v_{n-1}Rv_n$, where Π stops in state v_n,

$$\sum_{\delta(v_i,A_1t_1A_3\odot A_1t_1A_4\odot A_2t_1A_3\odot A_2t_1A_4)} 1 > \sum_{\delta(v_i,rt_1A_3\odot rt_1A_4)} 1$$

for $1 \leq i \leq n$ where all the involved basic problems are enabled. Determine a good firing rate for $A_1t_1A_3$, $A_1t_1A_4$, $A_2t_1A_3$ and $A_2t_1A_4$ is an optimisation problem for $\Lambda(A_1t_1A_3 \odot A_1t_1A_4 \odot A_2t_1A_3 \odot A_2t_1A_4)$.

Lets concern the SPNs of Fig. 2 and Fig. 3. Taking a $\mathcal{DS}_3$ model $\mathcal{A} = \langle W, R_\pi, M, (\Pi, \Lambda), \delta, \mathbf{V}\rangle$ that denotes this whole situation, (Π, Λ) will be the SPN denoted by two agents as in the SPN of Fig. 3 corresponding to a gens A_1 and A_2 of Fig. 4 and two agents as in the SPN of Fig. 2 corresponding to agents A_3 and A_4, and the other aspects of the SPN in Fig. 4. The function Λ will guarantee that the clock of A_1 will be faster than the clock of A_2 due to the rate returned by each transition of each subnet (the part of the whole SPN that models each agent). In the same way, the failure rate noticed by agents A_3 and A_4 will be modelled in Λ.

Now we can verify if it is possible that an agent interrupts its processing due to energy fault while processing some data (i.e. there are no tokens in the place Energy_i, where $i \in \{1, 2, 3, 4\}$ denotes the agent of which we are referring). We can also compute the probability that some processing leads to an error and the data will need to be parsed and processed again using lemma 1. In a state $w \in W$ where $s = M(w)$ it is equivalent to compute $\Pr(\langle s, \text{Verify}_3\text{Energy}_3 t_2 \text{Parse}_3 \odot \text{Verify}_4\text{Energy}_4 t_2 \text{Parse}_4\rangle \top \mid \delta(w, \Pi))$.

6 Conclusions

Usage of Petri Nets to model multi-agent systems is present in a large body of literature. We presented how model an agent in a Stochastic Petri Net and the usage of the logic $\mathcal{DS}_3$ may be applied to verify properties in these systems. The presented agent model can be extended and adapted to other agent approaches, encapsulating part of the complexity or detailing each step of computation.

The $\mathcal{DS}_3$ logic regards an extension of the Petri-PDL logic [17] (a Dynamic Logic that deals only with Ordinary Marked Petri Nets, where all transitions enabled have the same probability of firing). For scenarios in which the controlling of time (i.e. firing rate) is not required we do recommend to the reader the usage of this simpler logic. The choice by Stochastic Petri Nets instead of the ordinary ones leads to in increased expressive power where the user will be able to model scenarios more precisely. The complexity of the SAT problem also keeps in the same class (EXPTime-hard) [16].

As it can be seen in Section 3, the problem of verifying some properties is reduced to simple computations. The usage of $\mathcal{DS}_3$ leads to certify properties and compute the probability their insurance. Nevertheless the user is able to model using the graphical notion inherited from Petri Nets.

7 Further Work

Further work includes the extension of the Resolution calculus for Petri-PDL [26] to $\mathcal{DS}_3$ and the implementation of an automatic theorem prover. We are also working into an automatic model checking process with $\mathcal{DS}_3$.

References

1. Wooldridge, M.: An Introduction to MultiAgent Systems. John Wiley & Sons (2002)
2. Khosravifar, S.: Modeling multi agent communication activities with Petri Nets. International Journal of Information and Education Technology 3(3), 310–314 (2013)
3. Van Belle, W., Nose, K.: Agent mobility and reification of computational state: An experiment in migration. In: Wagner, T.A., Rana, O.F. (eds.) AA-WS 2000. LNCS (LNAI), vol. 1887, pp. 166–173. Springer, Heidelberg (2001)
4. Ganty, P., Majumdar, R.: Algorithmic verification of asynchronous programs. ACM Transactions on Programming Languages and Systems 34(1), 1–48 (2012)
5. Marsan, M.A.: Stochastic Petri Nets: An elementary introduction. In: Rozenberg, G. (ed.) APN 1989. LNCS, vol. 424, pp. 1–29. Springer, Heidelberg (1990)
6. Everdij, M.H.C., Klompstra, M.B., Blom, H.A.P., Klein Obbink, B.: Compositional specification of a multi-agent system by stochastically and dynamically coloured petri nets. In: Blom, H., Lygeros, J. (eds.) Stochastic Hybrid Systems. Lecture Notes in Control and Information Science, vol. 337, pp. 325–350. Springer, Heidelberg (2006)
7. Holvoet, T.: Agents and petri nets. Petri Net Newsletter 49 (1995)
8. Perše, M., Kristan, M., Perš, J., Kovacic, S.: Recognition of multi-agent activities with Petri Nets. In: 17th International Electrotechnical and Computer Science Conference, pp. 217–220 (2008)
9. Rongier, P., Liégeois, A.: Analysis and prediction of the behavior of one class of multiple foraging robots with the help of stochastic petri nets. In: Proceedings of 1999 IEEE International Conference on Systems, Man, and Cybernetics, EEE SMC 1999, vol. 5, pp. 143–148 (1999)
10. Haas, P.J.: Stochastic Petri nets: modelling, stability, simulation. Springer (2002)
11. Lyon, D.: Using stochastic petri nets for real-time nth-order stochastic composition. Computer Music Journal 19(4), 13–22 (1995)
12. Marsan, M.A.: Stochastic petri nets: an elementary introduction. In: Miola, A. (ed.) DISCO 1990. LNCS, vol. 429, pp. 1–29. Springer, Heidelberg (1990)
13. Marsan, M.A., Chiola, G.: On Petri Nets with deterministic and exponentially distributed firing times. In: Rozenberg, G. (ed.) Advances in Petri Nets 1987. LNCS, vol. 266, pp. 132–145. Springer, Heidelberg (1987)
14. Mazigh, B., Abbas-Turki, A.: Specifying and Verifying Holonic Multi-Agent Systems Using Stochastic Petri Net and Object-Z: Application to Industrial Maintenance Organizations. In: Petri Nets – Manufacturing and Computer Science. InTech (2012)
15. Babczyński, T., Magott, J.: PERT based approach to performance analysis of multi–agent systems. In: Rutkowski, L., Tadeusiewicz, R., Zadeh, L.A., Żurada, J.M. (eds.) ICAISC 2006. LNCS (LNAI), vol. 4029, pp. 1040–1049. Springer, Heidelberg (2006)
16. Lopes, B., Benevides, M., Haeusler, E.H.: Extending Propositional Dynamic Logic for Petri Nets. Electronic Notes in Theoretical Computer Science 305(11), 67–83 (2014)
17. Lopes, B., Benevides, M., Haeusler, E.H.: Propositional dynamic logic for petri nets. Logic Journal of the IGPL 22(5) (2014)
18. Feldman, Y.A.: A decidable propositional probabilistic dynamic logic. In: Proceedings of the Fifteenth Annual ACM Symposium on Theory of Computing, STOC 1983, pp. 298–309. ACM (1983)

19. Feldman, Y.A., Harel, D.: A probabilistic dynamic logic. Journal of Computer and System Sciences 28(2), 193–215 (1984)
20. Kozen, D.: A probabilistic pdl. In: Proceedings of the Fifteenth Annual ACM Symposium on Theory of Computing, STOC 1983, pp. 291–297. ACM (1983)
21. Tiomkin, M., Makowsky, J.: Propositional dynamic logic with local assignment. Theoretical Computer Science 36, 71–87 (1985)
22. Tiomkin, M., Makowsky, J.: Decidability of finite probabilistic propositional dynamic logics. Information and Computation 94(2), 180–203 (1991)
23. Lopes, B., Benevides, M., Haeusler, E.H.: Verifying properties in multi-agent systems using Stochastic Petri Nets and Propositional Dynamic Logic. In: Annals of the X Encontro Nacional de Inteligência Artificial e Computacional (2013)
24. Pujari, S., Mukhopadhyay, S.: Petri Net: A tool for modeling and analyze multi-agent oriented systems. International Journal of Intelligent Systems and Applications 10, 103–112 (2012)
25. de Almeida, E.S., Haeusler, E.H.: Proving properties in ordinary Petri Nets using LoRes logical language. Petri Net Newsletter 57, 23–36 (1999)
26. Nalon, C., Lopes, B., Haeusler, E.H., Dowek, G.: A calculus for automatic verification of Petri Nets based on Resolution and Dynamic Logics. Electronic Notes in Theoretical Computer Science (in press)

Multi-agent Coordination for a Composite Web Service

Anouer Bennajeh and Hela Hachicha

Stratégies d'Optimisation et Informatique intelligentE (SOIE),
41 Street of Liberty Bouchoucha-City CP-2000 Bardo – Tunis, Tunisia
anouer.bennajeh@gmail.com,
hela.hachicha@fsegs.rnu.tn

Abstract. The web services composition is one of the most important approaches ensuring to exploit several services by a single, called a composite service. The work presented in this paper proposes a web services composition model in order to reduce the network problems during the composition and reduce the total duration of composition. The proposed model is based on the coordination of software agents by exploiting its characteristics, hence it eliminate the passivity of web technology and make it more responsive. More than this model ensures the execution of various activities to a composite service, by the integration of an execution control strategy which offers solutions at real-time in case to execution problem from one of the composite service activities. Also, the execution control strategy reduces the response time to a web service composition request and reduces the overhead of network traffic.

Keywords: web service composition, multi-agent system, mobile agent.

1 Introduction

The evolution of Internet and the competitiveness between companies contributes to the appearance of several new web approaches. The web service composition is one of the new approaches which enrich the company activities by other external services, in order to meet to the needs of their clients, to reduce time and costs, and to increase overall efficiency in businesses.

The workflow approach appears as the first work that provides the composition and that is based on two modeling. Firstly, the orchestration modeling which is a set of actions to be performed by the intermediate web services [1]. The orchestration is to program a motor from a predefined process by the intermediate web services [5]. This modeling is based on a centralized vision. The second modeling is the choreography, which implements a set of services without any intermediate in order to accomplish a goal [1]. This modeling allows the design of decentralized coordination applications.

There are some works which improve the workflow approach, among these works; we cite the approach of [4] which uses the semantic approach to make the code understandable to the machines, and then permitting the automation of web services selection. However, according to [8] the web service composition approach is still has

J. Bajo et al. (eds.), *Trends in Prac. Appl. of Agents, Multi-Agent Sys. and Sustainability*,
Advances in Intelligent Systems and Computing 372, DOI: 10.1007/978-3-319-19629-9_10

many problems coming from web technology itself, such as: the absence of information on the web service environment and the passivity of this technology. Moreover, the web service composition can encounter execution problems of one of its services after the selection, whereas this problem influences on the response time and network traffic overload.

In the first place, the problem proposed by [8] can be treated with the apparition of software agent technology which provides many solutions due to their very innovative characteristics, such as: autonomy, intelligence, reactivity, collaboration, cooperation, mobility, communication, etc. Also, software agent technology can achieve their goals with a flexible manner by using a local interaction and/or remote with other agents on the network. In the second place, the execution problem of a composite service requires following the execution of various activities after the selection of web services.

In this context, we are interested to define and implement a model ensuring the composition of web services based upon on multi-agent coordination. We propose essentially to treat the two problem of response time and the network traffic overload. Furthermore, this model ensures the integration of an execution control strategy which offers solutions to real-time.

This paper is structured as follows. Section 2 presents the related work. In section 3, we describe our model for web services composition. Section 4 presents the rules of interaction between software agents. In section 5, we show the evaluation part of our model. Finally section 6 summarizes the paper.

2 Related Work

Recently the web services composition based on software agent technology has been experiencing much attention. These approaches are classified in two classes. The first class is based on multi-agent system. In [3] authors have proposed a multi-agent system for web services composition based on competition between coalitions of services. In this works, each service is represented by a software agent who contacts other agents and offers their services according to their reasoning ability. Then, the different coalitions will make the most competitive offer, where each solution receives a note, and the solution with the highest score will be selected. Also work in [7] has proposed a multi-agent system for web services composition. By contrast, it ensures the formation of only coalition by adopting the negotiation mechanism.

The second class of approach is based on mobile agent. In this context, authors in [9] ensure the composition of web services with the bottom-up approach, where the selection step occurred before the creation of activities plan. Unlike authors in [3], the selection of the web services occurred after the preparation of activities plan.

Authors in [9] seek to reduce the intermediary activities of a composite service, in order to reduce response time of composite service. In [6] authors have integrated a strategy to control the execution of the activities plan, but they did not offer an immediate solution when a problem of execution of any service was occurred. All these approaches previously described are useful and interesting contributions that have provided solutions at the composition of web services, by taking advantage of features offered by the software agent technology. However, some problems are not treated well by these works, such as, the response time of web service composition. In addition, some approaches do not proposed to control the execution of different components of a composite service, which can causes some network problems, such as, network overload, security. Also, these approaches have not proposed an immediate solution to an execution problem. In this sense and in order to improve the web service composition, we propose a new model of web service composition based on multi-agent system and mobile agent technologies.

3 Proposed Web Service Composition Model

The proposed web service composition model is based on the coordination between three agents: User Agent (UA), Service Agent (SA) and Compositor Agent (CA). In the Figure 1, we present the architecture of our model.

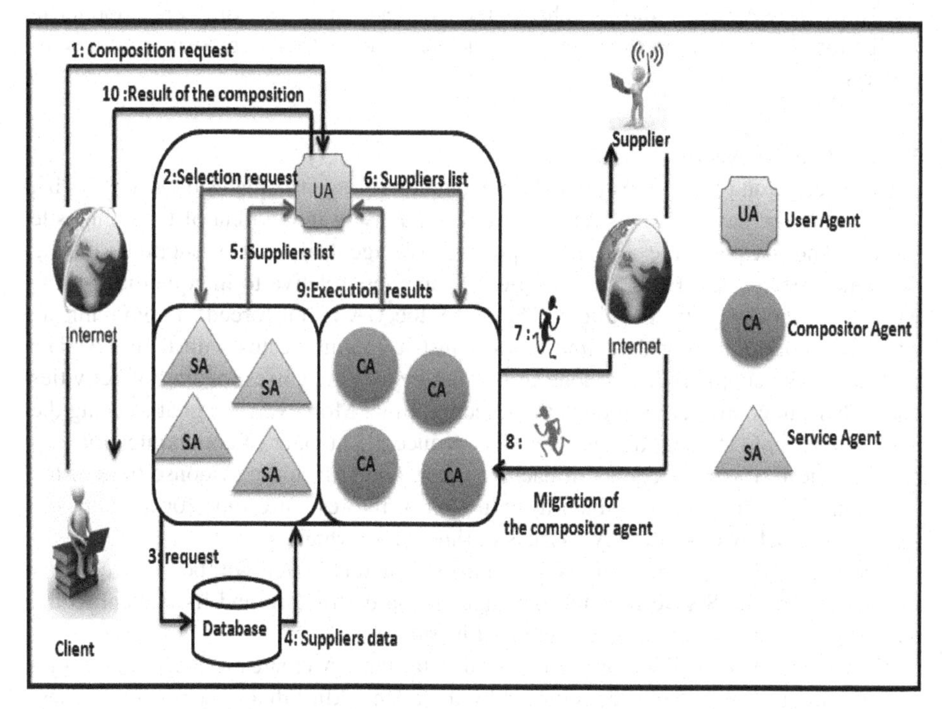

Fig. 1. The proposed architecture of our web services composition model

The proposed model is based on two functionalities. In the following sub-sections, we will detail the two functionalities.

3.1 Creation of the Activities Plan and the Suppliers List

This functionality is based on the UA and the SA.

3.1.1 User Agent

The role of UA is like the motor component of the orchestration model in the work of [5]. UA is a stationary agent, which communicates with SAs and CAs and receives the client's composition requests. The behavior of the UA in this functionality consists of three tasks which are:

Formation of activities plan: In the first task, the UA starts to verify the composition request sent by the client and extracts the web services and the data from the composition request in order to create the activities plan for a composite service.

Classification of the data: The second task is to check the activities plan and to classify the data sent by the client according to their correspondence to the services in the activities plan. These data are the input parameters for each web service requested.

Validation of the suppliers list: The third task starts after the creation of the suppliers list by the SA, where the UA checks the existence of at least only one supplier in the list for each activity in the plan, in order to ensure the total composition of a composite service.

3.1.2 Service Agent

After the creation of activities plan, the SA (stationary agent) creates the suppliers list by selecting the best suppliers matching to each activity in the plan of the composite service. The SA plays the role of a representative agent of a set of suppliers having the same web service. Each SA classifies its suppliers relative to the qualities of service (QoS) requested by the client. Therefore, the UA is not forced to communicate with all suppliers having the same service, just it communicates with their SA. The creation of the suppliers list is done in parallel by all SAs corresponding of activities plan. This can contribute to reduce the selection time. Moreover, the goal to using the SA is to facilitate the negotiation operation, reduce the number of communication acts between the UA and the agents of each supplier, and reduce the response time of the selection step. This can contribute to reduction of the response time for a composite service. The behavior of the SA consists of three tasks which are:

Validation: Each SA has a message standard structure which will be received from the UA. Then, the SA verifies the message standard structure and its content to the parameters compatibility of input/output of its service.

Extraction: After validating the received data, the SA creates a query taking into account the qualities of service required by the client. After that, it sends this request to the database which contains all information about the suppliers having its service. The SA agent is the only agent which has the right to access in the database of the composition system to select the best suppliers.

Classification: Finally, the SA classifies the suppliers taking into account their qualities of service (QoS), where it specifies the primary supplier and the reserve suppliers which can replace the primary supplier when it has an execution problem.

In our model the proposed of database is inspired from the work of [2]. In fact, the database of the composition system plays the role of the UDDI web service directory, where it contains the data about the qualities of service for each supplier. In this manner, the selection of web services corresponding to the activities plan is done in record time, because the selection of web services is performed locally. This is help to avoid to discover web services on the UDDI web service directory and to select the best suppliers by using a remote method. In other words, with discovered and selection of web services locally, we can avoid several problems such as: the security of each remote request, the overload of traffic on the network and the variation of the response time of each query which depend on the state server to visit and the state of network traffic.

3.2 Execution Control Strategy

The second functionality ensures the execution control strategy of different activities of the composite service which was created in the first functionality. The execution of this functionality is based on the UA and the CA.

3.2.1 User Agent

In the second functionality, the UA reaches two tasks which are:

Assignment of the suppliers list: The UA assigns each suppliers list to a CA.

Confirmation: The last task that the UA will do it is to check the result sent by the CAs. If this result contents an execution confirmation of one of its suppliers list, then the UA can send the result of the composition to the client.

3.2.2 Compositor Agent

The CA is a mobile agent; which responsible to ensure the execution of one service in the suppliers list. In fact, the CA migrates to the supplier agent platform and ensures the service execution according to the requirements of its client. The CA reaches two tasks which are:

Migration: After the reception of a suppliers list, the CA migrates to the first supplier agent platform in its list in order to ensure the execution of its service. If the CA encounters an execution problem (the web service server or the supplier agent platform is down), then the CA migrates to the next supplier in its list. The CA repeats the same steps until it has an execution confirmation by one supplier of its suppliers list.

Communication with Supplier agent: At its arrival, the CA communicates with the supplier agent in order to execute its service according to the client requirements.

Therefore, with the defined execution control strategy, we propose an immediate solution when an execution problem is encountered. This is due to the use of mobile agent (Compositor Agent). Thus, our model contributes to reduce the response time of the composite service and to reduce the traffic on the network.

4 Rules of Interaction between Software Agents

4.1 Rules for User Agent

During interaction with the SA, the UA sends a message containing an object composed of two elements. The first presents ID of SA, where each SA has an ID. Thus with this element, SA checks if this message concerns it or not. The second element presents the customers' requirements. In fact, each message concerns only one service of composite service.

$$\text{Required\_Service(ID\_Agent\_Service,Data[i])} \tag{1}$$

Where i = [1 to n], n is the total number of data.

During the interaction with CA, the UA sends a message containing an object composed of two elements. The first element presents the data which the supplier needed to respond to client's requirement. The second element presents the suppliers list containing the main supplier with the reserve suppliers. In fact, the first element will be used by supplier agent. The second element will be used by the CA to solve the execution problem with the main supplier.

$$\text{List\_Assignment (data[i], supplier[j])} \tag{2}$$

Where i = [1 to n] and j = [1to k], k presents the total number of suppliers which can respond to the customer's requirement.

4.2 Rule for Service Agent

During interaction with the UA, the SA sends a message containing a suppliers list organized according to their quality of service.

$$\text{Suppliers\_List(Supplier[i])} \tag{3}$$

Where i = [1 to n], n presents the total number of suppliers which can respond to the costumer's requirement.

4.3 Rules for the Compositor Agent

During interaction with the UA, the CA sends a message containing an object composed by two elements. The first element presents the response to execution of main service, where this answer is in the Boolean form (accepted or not accepted). The second element presents the results after execution of the service required.

$$\text{Response\_to\_execution (Execution, Result[i])} \tag{4}$$

Where Execution = accepted / not accepted and i = [1 to n], n is the total number of result. If Execution = not accepted then the Result = Φ.

5 Evaluation

In our model, we have proposed a solution to reduce the composition time and we have proposed an execution control strategy of the composite service. As a consequence, these new proposals allow to improving the response time to the composite service, to reducing the overhead of network traffic and to reducing the execution time of the composite service. In order to evaluate our model, we have developed a web application for a travel agency that ensures the composition of different services, which are: the hotel booking, the car rental and the purchase of a flight ticket. Firstly, the customer fills out a form which contains his requirements for each service. Secondly, this application performed a research to the best suppliers for each service required, executes each service according to the requirements of its client and sends the final response. In our travel agency web application, with a click, the customer can benefit to three services at a same time. This allows reducing the time of research of the best supplier and executing each web service in the best time. For the implementation of the software agents (UA, SA, CA) we have used the JADE platform.

Figure 2.a shows the durations of different stages of the construction and the execution of a composite service based on our model. The diagram "B" presents the total duration of composition (the creation of the activities plan for a composite service and the creation of the suppliers list for each activity), which was done locally in 1916.66 milliseconds. The diagram "C" presents the duration of best suppliers' selection, which was done in 1841.66 milliseconds.

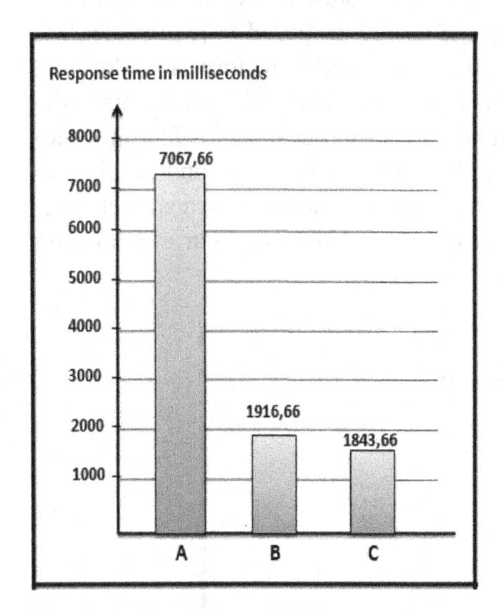

Fig. 2. a. Durations of different stages to construct and execute a composite service without execution problems

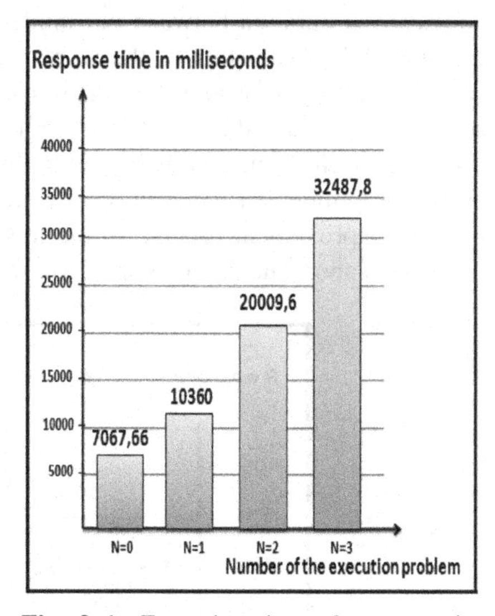

Fig. 3. b. Execution time of a composite service with variation of execution problems number

The diagram "A" presents the execution time of a composite service based on different services which was done in 7067,66. In fact, several factors influence on this step, such as the rate of traffic on the network, the server status of each service requested. We have tested our execution control strategy on a local server.

In Figure 2.a, we present an evaluation case without taking into account the execution problems. For this reason, in the second evaluation, we varied in each time the number of the execution problems for each activity. In contrast, in Figure 2.b, we present the execution times of a composite service with variation to the number of the execution problem that the CA can to encounter when it migrates over the Network. We notice that the variation of execution problems number influence the total execution time of a composite service. The resolution of these problems is realized by our execution control strategy with the proposal of a solution at real-time, where every time the CA encounters an execution problem with first supplier; it changes its destination by taking the reserve supplier without returning to its starting platform. So, the loss average with each execution problem is 4112,18 milliseconds which is an acceptable period, compared to repeating the step of selecting and executing a new supplier.

In the literature, the web service composition based in two fashions, which are: a composition in a top-down fashion and a composition in a bottom-up fashion. In our work we adopted a top-down fashion and this is the same thing with the web service composition model of [3]. Thus, with this fashion, we need to start with the creation of an activities plan for each composition request. In this case, [3] adopted the coalition model to ensure the coordination between agents to create the activities plan. However, in our work we adopted the hierarchical model, where, only the UA is responsible to create the activities plan. In Figure 3 we present a comparison of the composition duration between our model and the model proposed in [3]. In fact, according to the evaluation of the work in [3], the best duration to form a coalition is 3000 milliseconds, where the coalition reflects to an activities plan of a composite service. In contrast, the better duration to composition with our model is 359 milliseconds, where this duration includes the time to form an activities plan and the time to create a suppliers list for each service. We notice that our model presents a very significant improvement relative to the model proposed in [3], where our composition duration is about the tenth compared to the duration of [3].

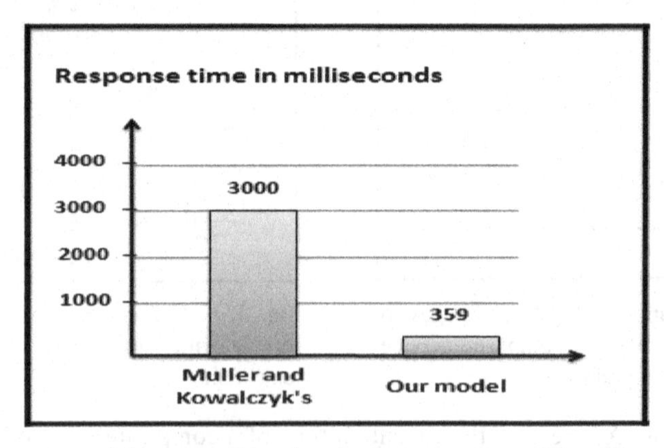

Fig. 4. Comparison of the composition duration between our model and the model of [3]

6 Conclusion

The aim of this paper is to address essentially solutions of three problems: the response time, the overload on the network and the execution problem of different activities of a composite service. We solved these problems by the creation of a web service composition model based on a coordination of software agents. The proposed model is composed by two main functionalities. The first functionality ensures the creation of activities plan for the composite service and the selection to the best suppliers which can respond to different activities. These two stages are realized locally, on the same server of the composite service. The second functionality ensures the execute control which offers the solutions at real-time in case where a composite service activity has an execution problem.

References

1. Benatallah, B., Casati, F., Grigori, D., Motahari-Nezhad, H.R., Toumani, F.: Developing Adapters for Web Services Integration. In: Proceedings of the 17th International Conference on Advanced Information System Engineering, Porto, Portugal, pp. 415–429 (2005)
2. Kouki, J., Chainbi, W., Ghedira, K.: Local Registry-Based Approach for Binding Optimization of Web Services. International Journal: The Systemics and Informatics World Network 11, 69–76 (2010)
3. Muller, I., Kowalczyk, R.: Towards Agent-based Coalition Formation for Service Composition. In: Proceedings of the IEEE/WIC/ACM International Conference on Intelligent Agent Technology (2006)
4. Osman, T., Thakker, D., Al-Dabass, D.: Bridging the gap between workflow and semantic-based web services composition. In: Proceedings of WWW Service Composition with Semantic Web Services, Compiègne, France, p. 1323 (2005)
5. PELTZ, C.: Web Service Orchestration and Choreography. A look at WSCI and BPEL4WS. Web Services – HP Dev Resource Central (2003)
6. Qian, Z., Lu, S., Xie, L.: Mobile-Agent-Based Web Service Composition. This work is partially supported by the National Natural Science Foundation of China under Grant No.60402027 and the National Basic Research Program of China (973) under Grant No.2002CB312002 (2005)
7. Qian, W.: A web services composition model for coalition net construction. In: International Conference on Computer Science and Service System (CSSS) (2011)
8. Tong, H., Cao, J., Zhang, S., Mou, Y.: A fuzzy evaluation system for web services selection using extended QoS model. Library Science: Journal of Systems and Information Technology- Table of Contents 11(1-2) (2009)
9. Zhang, Y.: Web Service Composition Based on Mobile Agent. In: International Conference on Internet Computing and Information Services IEEE (2011)

Achieving Parkinson's Disease Patient Autonomy through Regulative Norms

Jorge J. Gomez-Sanz and Pablo Campillo Sanchez

Universidad Complutense de Madrid, Madrid, Spain
{jjgomez,pabcampi}@ucm.es

Abstract. In an Ambient Intelligence system for humans, devices in the environment perform actions to improve the quality of the daily living of a person. Research in normative systems could be applied to regulate the operation of this kind of systems. Nevertheless, determining the right norms is not trivial. This contribution discusses the case of a Parkinson's Disease patient. In particular, the paper studies a TV watching scenario where the right moment to assist the patient is the subject of discussion. Results are obtained from the SociAAL project, whose aim is the integration of social sciences concerns into the development of this kind of systems and the reduction of their production cost.

Keywords: Ambient Intelligence, Ambient Assisted Living, Autonomy, Parkinson Disease, Modeling, Simulation, Normative system, Norm.

1 Introduction

Ambient Intelligence (AmI) deals with the interplay of actors, either human or devices, while pertaining to, or moving from, a set of known contexts. This paper focuses on applying norm research to regulate that interplay by means of a normative system. A normative system intend to preserve some desirable system dynamics by rewarding/punishing behaviors. Also, they specify how and to which extent actors can modify the norms [2].

There is preliminary work in norms applied for Multi-Agent System (MAS) in AmI [3], but its scope is conceptual, yet. To facilitate advances in this direction, simulators for AmI can be a key element. Simulators like UbikSim [12] or PHAT [4] permit to create more complete scenarios that involve the human actors, how they use the AmI elements, and how these elements may return feedback the humans.

As an example, the paper discusses autonomy concerns of a Parkinson's Disease (PD) patient in a TV watching scenario. This case is analyzed to determine how the AmI system is aiding and how a normative approach could drive the AmI behavior. The complete case is not included in this paper. The paper addresses mainly how such normative system could be deployed and how it would relate to the autonomy preservation criteria. A non-experienced developer may think the more the system does for its user, the better. In this paper, we define the autonomy preservation of the human as the capability of the human to do things without assistance (artificial aid included). This

J. Bajo et al. (eds.), *Trends in Prac. Appl. of Agents, Multi-Agent Sys. and Sustainability*,
Advances in Intelligent Systems and Computing 372, DOI: 10.1007/978-3-319-19629-9_11

autonomy may conflict with the safety of the PD patient. A solution to the dilemma would be increasing the assistance to the patient as the risk of imminent injuries raises.

The *Social Ambient Assisted Living* (SociAAL) project [9] proposes including social sciences concerns in the definition of AmI systems for people with Parkinson's Disease (PD) and the development cost reduction of such systems. SociAAL project has produced the PHAT [4] simulator that, in our opinion, offers new opportunities for defining and performing experiments in AmI. It permits to re-create scenarios involving humans into computer simulations using a hardware/software-in-the-loop approach. A developer can provide a description of a house/room, which devices exist inside, simulate those devices or connect real devices to the simulated environment (hardware in the loop), or develop software for the control of those devices within the simulation in a reusable way (software in the loop).

The contribution is organized as follows. Section 2 contains some thoughts about the PD and how it affects the autonomy of the patients. There is also a discussion of the scope of a simulation that captures the TV watching scenario. Section 3 is dedicated to explain the deployment of the normative system and what it would do. Section 4 discusses the connection between this contribution and other related ones. Finally, Section 5 contains the conclusions.

2 Autonomy of Parkinson's Disease Patients in a TV Watching Scenario

A person is the more autonomous the less assistance from other humans is needed, but this is not an absolute. Common daily activities are not always affordable for PD patients. However, they have to insist and not just let anyone/anything do everything everytime for them. So, even if they are slow or less effective, it is more positive for them to keep trying if they really can success ... unless there are safety concerns. From this perspective, it is necessary to devise ways in which an assisting AmI system can check and validate/facilitate activities.

Such ways are recommendations that devices in the AmI may or not follow. However, a normative system could realize this kind of knowledge, formalizing which kind of collaboration is desirable or unacceptable, informing incoming devices of the existence of such norms, and enabling, perhaps, the devices to modify existing norms [2].

To illustrate the utility of normative evaluation of the system dynamics, a scenario involving humans and devices is built using PHAT and its modeling language, called SociAAL. What the scenario evaluates is the control software inside of these devices, which are expected to be software agents.

In the scenario, the PD patient goes to the living room and wants to watch TV. To correctly assess the performance of the system and how it affects the autonomy of the PD patient, the original scenario needs to be broken down and parametrized. The actions are modeled in figure 1 and rendered into behaviors in a 3D environment with PHAT. The action sequence is *Go to the living room*, which makes the patient move from its current location to the living room; *switch on the TV*, which consists on using the remote control to turn on the TV (depending on degree of tremors,the patient may press several buttons, only one, or none); *switching channels and watch them*, where the patient presses a concrete channel button (the action may be affected by tremors and more than one button, only one,or none be pressed); and *switch off the TV* with the remote. Each action is further broken down to add details. Interested readers may consult the PHAT tutorial in `http://grasia.fdi.ucm.es/sociaal`.

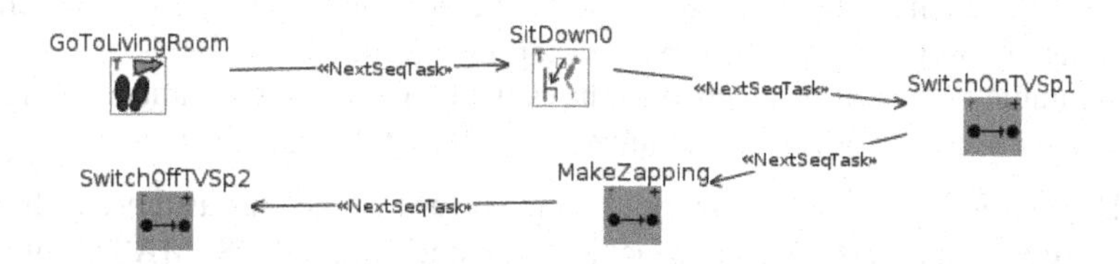

Fig. 1. Action sequence described with SociAAL modeling language

The PD disease effects on daily living activities are expressed with the modeling language too. The stage E1 of the PD is associated to tremors of three different degrees: low, medium, and high. The effect of these tremors over some activities is detailed in figure 2. The *go to living room* activity is performed slower, and the *switch on TV* is replaced by a command to ask for help. Button pressing is equally affected by modifying the touch accuracy.

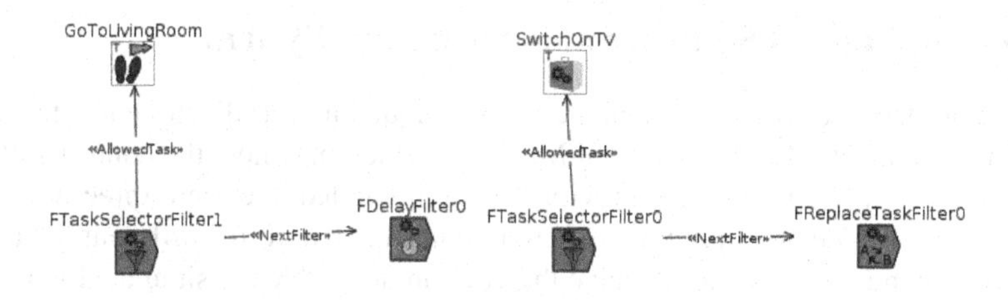

Fig. 2. Medium tremors effect on basic activities from figure 1

Ideally, the PD patient should be able to go through these activities without assistance. Depending on the degree of tremors, some actions may be harder or just impossible. Among others, the following situations are considered: power-on button of the TV may not be successfully pressed either on the TV or on the remote control; channel switching may not work because several buttons are pressed at the same time; dropping the remote control because of a defective grabbing; the patient may fall while sitting down or while walking; or, while walking too, the patient may face a frozen feet situation where the patient seems unable to keep the movement.

From these events, the paper focuses on the button (power on and channel ones) pressing problem when the patient has tremors. This activity is monitored by the devices inhabiting the target AmI system, as shown in figure 4:

– An intelligent remote control. The remote control can control a capacitive area that represents the buttons. It can provide measures of which buttons are been pressed, even if there are many at the same time. It also can modify the size of the buttons in the screen to make them bigger if the patient has tremors. It can detect too when the device is on the ground and take measures to let someone else to retake it. Layout rearrangement is possible too.

- A smart TV. The TV has capacitive controls too. These controls permit to turn the TV on/off, change channels, and modify the volume. It also have a camera that permits to observe the user and interact through gestures.
- A wearable device. It is attached to the patient's wrist. This wearable can interact with the different devices to customize the experience of the user.

A 3D scenario is created with dummy users performing actions as those of figure 1. Devices are emulated with Virtual Machines running Android OS. MAS software, or equivalent, is expected to be hosted within those Virtual Machines. Their actions are the ones to be under normative control.

The interaction between the patient and devices, and among devices, is possible. Between the patient and the device, there are several possibilities. Android foresees a number of actions: touch the capacitive display, press buttons, speak to it, listen to sounds, perceive proximity, produce vibration, perceive movements of the device, and so on. Inter-device cooperation is an interoperability issue that is not addressed. It is assumed devices can already cooperate in order to deal with the patient's activities. The complete video of the simulation can be consulted at the demo section of the project SociAAL `http://grasia.fdi.ucm.es/sociaal`.

3 Adding Control Software and Normative System

The simulation of the scenario is deployed in a computer host as illustrated by the figure 3. The devices are Android OS Virtual Machines which may host the control software for the device they represent. As previous section detailed, there are three devices to consider and each device may host different control software for different situations. The devices sends actions and perceive the environment which is simulated with a 3D representation. Inside the 3D simulation, actions are observed and evaluated by the normative system.

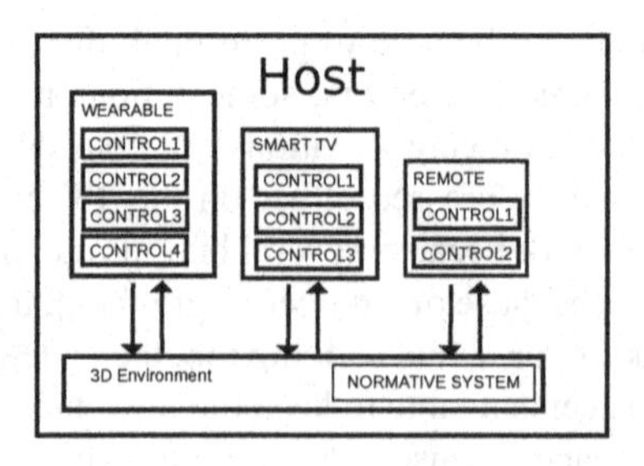

Fig. 3. The normative system and the other AAL devices

The resulting scenario is shown in the figure 4. The simulation depicts the remote control (middle bottom), a smart TV abstraction (bottom left), the 3D environment (right side), a wearable device (middle top), and the JADE platform keeping references to existing agents within the devices (left top corner). The user is expected to use the remote to change channels and the actions are processed by the agents within. The agents in the remote communicate with agents in the smart TV to switch channels or interchange other information. For the sake of clarity, instead of showing images, text is used to distinguish clearly the current channel. Actions over the remote are performed using the tactile display and taking into account the patient tremor status.

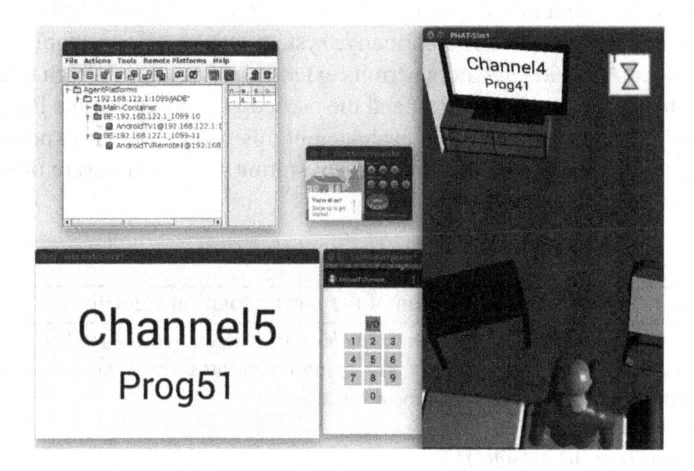

Fig. 4. Simulation scenario with android devices running JADE agents

As actors in the simulation are dummy users, their status can quantified and its reactions assessed. It is possible to define what a desirable autonomy means for this dummy and qualify attempts to interfere as negative. This is formalized as the *autonomy preservation criteria* and realized by means of the normative system:

- *Undesirable interruption*. If the action is going to be successful and it is interrupted, the interrupting device is punished. Interruption is understood as some device generating unexpected stimulus.
- *Tolerated failure*. If the action is not going to be successful and the devices do not assist the patient, they are punished.

The normative system has to enforce this *autonomy preservation criteria* by means of the two aforementioned norms. The violation of the first norm *undesirable interruption* punishes those devices that assist the user too much. The violation of the *tolerated failure* norm punishes those not assisting when they should.

The normative system development is possible because there is perfect knowledge of the environment (what is happening and what is going to happen), and every element in the simulation can be fully inspected. For the simulation, it means there is a perfect oracle giving perfect predictions.

Control software is deployed in the virtual devices just as any other Android application. An Android application can be developed in many ways. We assume the developer uses mainly *activities* and that invoke other *activities* through *intents*. Another alternative consists in defining APIs for control software to access the critical elements of the mobile. The important part is the capability to identify actions issued for the device itself or towards others.

Algorithm 1 resumes the way this normative system would work. Actions are obtained from each Virtual Machine representing a device and put together into a queue. Norm violation occurs if either there is an action when there shouldn't or if there is no action when there should. Punishment may not be immediate, if the system designer decides so. There may be initial notifications about the norm violations followed by final warnings about consequences. The normative system may even stop ongoing activities in a device, since Android emulators permits external intervention. Which one is better depends on the current AmI under test and the particular condition of the PD patient.

A concrete control software is not evaluated in this paper, but it is expected for future works. It is expected the behavior of the resulting AmI to be aware of the norms violations and react accordingly.

Algorithm 1. Preliminary description of the norm violation algorithm

Input: The two considered norms $\{undesirableInterruption, toleratedFailure\}$ that represent the autonomy criteria, $situation$ represents the current status of the dummy, $punish$ is the punishment function

```
1  while thereAreActions do
2      update(CurrentActions)
3      for action ∈ CurrentActions do
4          if violates(action, undesirableInterruption, situation) then
5              punish(responsible(action))
6
7          else
8              if violates(action, toleratedFailure, situation) then
9                  punish(responsible(action))
```

4 Related Work

Ambient Assisted Living work is usually done in experimental labs and, in most cases, using the Living Lab method. Fernandez-Pastor et al. [8] talk about a Virtual Lab in the sense of something that allows a developer to pretend be the patient. That concept is similar to the *participative simulation* one. The work introduced in this paper takes a different perspective. The developer does not participate, but observe the system dynamics expecting the dummies to trick the devices.

The concept of patient in this work is closer to the idea of *Virtual Actor*, but not as demanding. There is no intention of achieving human-to-computer interaction through different media, as in [1], or reproducing with realism the human movement, as in [5], or capturing the whole spectrum of human beings' capabilities [11]. The goal is to bring key aspects from the real world to a simulated environment using a limited amount of resources, enough realistic to convince devices that the world they are connected to is the real one. Patients in this simulation are dummies, but there is technology to make them more capable. Listening could be achieved through standard speech understanding software, as proposed in [10], where a number of technologies are reviewed to make characters more capable.

Cortes et al. [6] propose the idea of shared-autonomy where the assisting device adjust the support they provide. This capability is required here as well. The difference is the definition of autonomy itself. The PD patient autonomy has to be preserved and, in order to achieve this, violating behaviours are punished.

OVACARE [7] is a health care system for Alzheimer patients. It uses GORMAS as design methodology to represent different virtual organizations to capture the way the system is structured. In the system, norms are defined to ensure the behaviors within the organizations is the expected. Compared to this contribution, norms shown in [7] do not account for the clinical assessment of the patient as in the autonomy example. Also, this paper is not using any particular methodology.

Caire [3] defends the role of norms to regulate behaviors in the cities. The paper studies the idea of *conviviality*, which is a group where *individuals are welcome and feel at ease*. In this work, the conviviality concept matches our purpose. Devices and patients ought to be at ease with each other. The contract among them is represented by several norms that to be implemented in the normative system.

5 Conclusion

The dynamics of a Ambient Assisted Living system can be regulated using research borrowed from MAS. In particular, we aim to regulate the interaction between devices and one patient in an Ambient Intelligence scenario through regulative norms. The norm do refer to the recommendation that PD patients ought to keep their routines as long as they can perform them safely and by themselves. We translate this idea as the preserving autonomy criteria norms.

The work has been made within the SociAAL project, where the ultimate goal is reducing the development costs of AAL systems for people with PD and integrating social concerns into their development. The concept of autonomy preservation is a good example of social concern. By supporting the independence of the patient, their caregivers, usually relatives, are supported too.

The example assumes it is possible to have an oracle, something that foresees when the scenarios will be successful. This cannot be ensured in a real situation, though. Even if the oracle function is only valid during simulation, it still provides good starting points for creating and testing other possible normative solutions.

Future work will inspect this concept of autonomy when other humans get in the loop. Also with longer and more complex activities of the daily living. Current results permit to run hours of activities. However, the interaction between devices and the activities needs further work.

Acknowledgement. We acknowledge support from the project "SOCIAL AMBIENT ASSISTING LIVING - METHODS (SociAAL)", supported by Spanish Ministry for Economy and Competitiveness, with grant TIN2011-28335-C02-01 and by the Programa de Creación y Consolidación de Grupos de Investigación UCM-Banco Santander, call GR3/14, for the group number 921354 (GRASIA group).

References

1. Baldassarri, S., Cerezo, E., Seron, F.J.: Maxine: A platform for embodied animated agents. Computers & Graphics 32(4), 430–437 (2008)
2. Boella, G., van der Torre, L.W.N., Verhagen, H.: Introduction to normative multiagent systems. In: Normative Multi-agent Systems, 18.03. - 23.03.2007 (2007)
3. Caire, P.: A normative multi-agent systems approach to the use of conviviality for digital cities. In: Sichman, J.S., Padget, J., Ossowski, S., Noriega, P. (eds.) COIN 2007 Workshop. LNCS (LNAI), vol. 4870, pp. 245–260. Springer, Heidelberg (2008)
4. Campillo-Sanchez, P., Gómez-Sanz, J.J.: Agent based simulation for creating ambient assisted living solutions. In: Demazeau, Y., Zambonelli, F., Corchado, J.M., Bajo, J. (eds.) PAAMS 2014. LNCS (LNAI), vol. 8473, pp. 319–322. Springer, Heidelberg (2014)
5. Cavazza, M., Earnshaw, R., Magnenat-Thalmann, N., Thalmann, D.: Motion control of virtual humans. IEEE Computer Graphics and Applications 18(5), 24–31 (1998)
6. Cortés, U., Annicchiarico, R., Urdiales, C., Barrué, C., Martínez, A., Villar, A., Caltagirone, C.: Supported human autonomy for recovery and enhancement of cognitive and motor abilities using agent technologies. In: Agent Technology and e-Health, pp. 117–140. Springer (2008)
7. De Paz, J.F., Rodríguez, S., Bajo, J., Corchado, J.M., Corchado, E.S.: OVACARE: A multi-agent system for assistance and health care. In: Setchi, R., Jordanov, I., Howlett, R.J., Jain, L.C. (eds.) KES 2010, Part IV. LNCS (LNAI), vol. 6279, pp. 318–327. Springer, Heidelberg (2010)
8. Nieto-Hidalgo, M., Ferrández-Pastor, F.J., García-Chamizo, J.M., Flórez-Revuelta, F.: DAI Virtual Lab: a Virtual Laboratory for Testing Ambient Intelligence Digital Services. In: Proceeding of V Congreso Internacional de Diseo, Redes de Investigacin y Tecnologaa Para Todos, DRT4ALL (2013)
9. GRASIA. SociAAL project web site (2014)
10. Gratch, J., Rickel, J., André, E., Cassell, J., Petajan, E., Badler, N.: Creating interactive virtual humans: Some assembly required. Technical report, DTIC Document (2002)
11. Gratch, J., Rickel, J., André, E., Cassell, J., Petajan, E., Badler, N.I.: Creating interactive virtual humans: Some assembly required. IEEE Intelligent Systems 17(4), 54–63 (2002)
12. Serrano, E., Botia, J.A., Cadenas, J.M.: Construction and debugging of a multi-agent based simulation to study ambient intelligence applications. In: Cabestany, J., Sandoval, F., Prieto, A., Corchado, J.M. (eds.) IWANN 2009, Part I. LNCS, vol. 5517, pp. 1090–1097. Springer, Heidelberg (2009)

How Many Kinects Should Look At You? A Multi-Agent System Approach

Miguel Oliver[1], Francisco Montero[1,2], José Pascual Molina[1,2], Pascual González[1,2], and Antonio Fernández-Caballero[1,2]

[1] Universidad de Castilla-La Mancha, Instituto de Investigación en Informática de Albacete, 02071-Albacete, Spain
[2] Universidad de Castilla-La Mancha, Departamento de Sistemas Informáticos, 02071-Albacete, Spain

Abstract. This work compares the combined use of several infrared sensors such as the Microsoft Kinect. The system is modeled as a multi-agent system where each Kinect sensor is considered a physical/intelligent agent provided with pattern recognition capacity. The proposal offers a set of experiments that study when under concrete and different scenarios the combination of agent sensors achieves more accurate results than a single agent. So, the experiments analyze the accuracy and data resolution obtained from Microsoft Kinect sensors. Concretely, six experiments are conducted and the results are analyzed with different configurations of a variable number of Microsoft Kinect sensors. The results indicate that a single Kinect agent provides a higher accuracy in comparison to more Kinects. However, additional agents can be considered to increase the flexibility of the capture system.

Keywords: Kinect sensors, multi-agent system, capture precision.

1 Introduction

Currently, multi-agent systems (MAS) have been identified as one of the best suited technologies to contribute to the deployment of visual sensors, as they exhibit flexibility, robustness and autonomy (e.g. [1], [2]). Indeed, the MAS paradigm has been applied to computer vision in a number of different works (e.g. [3], [4]), including RGB-D cameras.

Microsoft (MS) Kinect is a camera-based sensor primarily used to directly control computer games through body movement. Kinect tracks the position of the limbs and body without the need of handhold controllers or force platforms. The use of a depth sensor allows Kinect to capture three-dimensional movement patterns. But, there are measurement errors in a depth image, called 'holes', which are due to different factors. Some of them are occlusions caused by objects or people, surfaces that absorb the infrared beam, disparity between transmitter and infrared (IR) receiver, saturation of outside IR light, and objects or people moving very fast in the recognition area. The problem of studying which is the optimal number of depth cameras to be installed in a setup is very important.

The data collected from a single Kinect does usually not present serious problems. Some recent studies (e.g. [5]) show that the error of the MS Kinect and Asus Xtion

J. Bajo et al. (eds.), *Trends in Prac. Appl. of Agents, Multi-Agent Sys. and Sustainability*,
Advances in Intelligent Systems and Computing 372, DOI: 10.1007/978-3-319-19629-9_12

sensor with distances of 2 meters is less than 2.5 centimeters. Another work shows that the error gotten with Kinect is too large for distances greater than 4 meters, but for shorter distances the error is acceptable [6]. An example of accuracy calculation is offered in a recently proposed rehabilitation system for the elderly [7]. In addition, a study of the rehabilitation of two people using a Kinect is shown [8]. Systems already implemented as TeKi, Toyra, KineLabs, Reflexion and VirtualRehab also support the idea.

The use of multiple sensors in one area of recognition is one of the points of most interest to researchers today. Indeed, solving the problem of the holes in depth images produced by occlusions will help to increase the maximum capacity of sensor recognition and to expand the range of action. For example, MS Kinect cameras are used to detect multiple users from different positions [9], thereby avoiding occlusions produced by the environment and occlusions produced by the body. There is also a study of increasing the accuracy of MS Kinect with the help of an RGB camera [10].

Increasing the number of sensors that point to a specific area could solve the problem of occlusions. But, the problem of saturation of infrared light is increased when using more than one sensor. A recent work [11] studies the noise produced by capturing the scene with two Kinect sensors simultaneously and provides some techniques to reduce it. Another work [12] shows the interference from up to three Kinect sensors that capture the same scene. Recently, a work [13] introduced an experiment with two Kinect sensors and five PlayStation Eye sensors. It is concluded that the accuracy is always less than 10 centimeters in both cases. In [14] also the issue of interference arising from two Kinect sensors when focusing on the same scene is addressed.

This article aims to complete the previously mentioned works by studying the problems encountered from IR saturation. A MAS model is used where each MS Kinect is considered a physical/intelligent agent. We study the problems that arise when using more than one Kinect sensor in the same area of recognition. This study focuses on how the distance to the subject and the number of sensors affect the localization of a user in the scene.

2 Description of the Experiments

The study described in this work consists of six experiments with three different Kinect configurations, where the number, location and the scope of the Kinect agents are modified. The configurations (*Conf1*, *Conf2* and *Conf3*) are associated to three different areas: 2×2, 3×3 and 4×4 square meters. A mannequin was located inside the scope area in each configuration and one hundred samples were collected at each measuring position, thereby achieving a total of 400 samples in the first configuration, 900 samples in the second configuration and 1600 samples in the third configuration, for each of the tests performed. Two metrics were especially considered: the accuracy of mannequin's position and the number of erroneous pixels in a frame. The number of erroneous pixels refers to positions of the depth image captured by the sensor, which cannot be obtained distance from this sensor. The use of a mannequin instead of a human is justified to avoid the movement that a real user involuntarily performs. A graphical representation of the experiments and configurations is shown in Fig. 1.

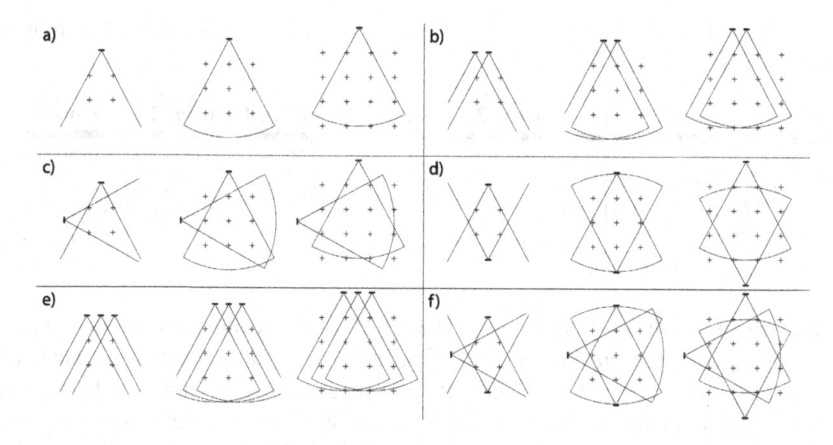

Fig. 1. Experiments and configurations. a) Experiment with one sensor. b) Experiment with two sensors pointing to the same direction. c) Experiment with two perpendicular sensors. d) Experiment with two confronted sensors. e) Experiment with three sensors pointing to the same direction. f) Experiment with three sensors separated 90 degrees among them.

Once the samples are collected, the results are processed in order to present them properly. The accuracy of the mannequin's position (in centimeters) is associated to the standard deviation of the data obtained by each one of the sensor agents. To achieve this, the position of the mannequin is calculated by arithmetic average of the data collected by each sensors, and from all points collected position standard deviation of these points is calculated. On the other side, the obtained erroneous pixels in a frame by the Kinect were documented by using their average.

2.1 Experiment #1: A Single Kinect in Front of the User

The first experiment is developed with only a single Kinect agent which collects data from a mannequin. The Kinect is located in front of the mannequin. The experiment is shown in Fig. 1a. These first values are used to establish reference values for posterior experiments. The results of the first experiment are shown in Table 1. The results of the first experiment perfectly show how the Kinect operates. When the distance is short, the standard deviation values are lower than 1 centimeter, being typically lower than 1 millimeter when the distance is equal to or lower than 2 meters. In distances of 4 meters (*Conf3*), these values are higher than 1 centimeter, reaching nearly 5 centimeters. In parallel, the erroneous pixels are higher than 50,000 pixels and rarely reach up to 55,000 pixels, so that these values can be considered a reference number for this metric.

2.2 Experiment #2: Two Kinect Sensors in Front of the User

This experiment is developed with two Kinect agents. The sensors are situated in front of the mannequin and between the agents there is a distance of 20 centimeters. This location of sensors is shown in Fig. 1b. The aim of this experiment is to analyze interferences or noise between two sensors with the same orientation. The results of this

Table 1. Standard deviation of mannequin's position accuracy (in centimeters) and erroneous pixels (in thousands) in experiment #1

Conf.1		**Conf.2**			**Conf.3**			
0,10	0,02	-	0,03	-	-	0,02	0,05	-
0,03	0,06	0,02	0,03	0,13	-	0,03	0,03	-
		0,12	0,08	0,18	0,22	0,13	0,12	0,36
					1,14	4,71	1,31	3,22

Conf.1		**Conf.2**			**Conf.3**			
58	43	-	53	-	-	63	46	-
51	53	51	49	49	-	54	55	-
		49	48	50	53	53	53	53
					52	53	53	53

second experiment are shown in Table 2. The results of the accuracy of the mannequin's position are worse than in the previous experiment #1. The worsening is especially relevant in the third configuration when the distance achieves 4 meters, where values between 3.44 centimeters and 5.51 centimeters are obtained. The behavior of the erroneous pixels in this experiment is lower than in the first experiment; the second rate of the erroneous pixels is better. We can conclude that erroneous pixels are directly proportional to the mannequin's position. A high erroneous frame rate means that there are recognition problems; however, a low number of erroneous pixels is not an indicator of better positioning.

Table 2. Standard deviation of mannequin's position accuracy (in centimeters) and erroneous pixels (in thousands) in experiment #2

Conf.1		**Conf.2**			**Conf.3**			
0,07	0,03	0,03	0,06	-	-	0,09	0,03	-
0,07	0,07	0,10	0,05	0,22	0,05	0,06	0,07	-
		0,15	0,21	0,22	0,21	0,14	0,22	0,25
					3,44	3,79	5,51	4,69

Conf.1		**Conf.2**			**Conf.3**			
52	46	47	46	-	-	64	42	-
49	46	49	49	52	41	51	47	-
		46	47	46	49	49	49	55
					48	49	49	48

2.3 Experiment #3: Two Kinect Sensors Located Perpendicularly

Two Kinects which are located perpendicularly are used in the third experiment. A Kinect is located in front of the mannequin and the other at an angle of 90 degrees (see Fig. 1c). The experiment is designed to determine how two Kinect sensors located perpendicularly work in comparison to the previous experiments. The results of this experiment are presented in Table 3. This experiment shows a deterioration in accuracy. The results on the right (last column) are greater than in experiment #1, because the mannequin is located 4 meters from the second Kinect. However, the erroneous frame rates are better than in previous experiments thanks to the presence of a second Kinect. We can emphasize that the erroneous frame rate and the accuracy of the mannequin's position are not directly proportional.

2.4 Experiment #4: Two Opposed Kinect Sensors

In the fourth experiment two Kinects are located one in front of the other. The location of the Kinects in this new experiment is shown in Fig. 1d. The purpose of this new

Table 3. Standard deviation of mannequin's position accuracy (in centimeters) and erroneous pixels (in thousands) in experiment #3

Conf.1	
0,04	0,14
0,31	0,04

Conf.2		
-	0,06	0,23
0,02	0,08	0,12
0,14	0,31	0,23

Conf.3			
-	0,02	0,26	4,72
0,55	0,05	0,32	2,04
0,14	0,26	0,10	0,52
1,02	6,25	2,01	2,05

Conf.1	
42	31
39	35

Conf.2		
-	42	44
49	41	40
49	39	39

Conf.3			
-	64	37	27
38	43	43	28
45	43	42	41
52	53	41	40

distribution of the Kinects is to analyze how an infrared beam interferes on the lens of another sensor agent. This disposition of Kinects makes their lasers hit each other. The results of this experiment are presented in Table 4. They show a more uniform distribution of values than in experiment #1 values for both mannequin's position and erroneous pixels. The average of the mannequin' position accuracy is worse than the results in experiment #1. However, the results of the erroneous pixels rate are lower than for experiment #1 in Conf1 and Conf2, but it is higher than for experiment #1 in Conf3. This distribution of Kinects presents interferences of the infrared beams and this issue affects the second metric.

Table 4. Standard deviation of mannequin's position accuracy (in centimeters) and erroneous pixels (in thousands) in experiment #4

Conf.1	
0,10	0,04
0,13	0,08

Conf.2		
0,27	0,10	0,66
0,08	0,02	0,45
0,11	0,10	0,17

Conf.3			
0,75	0,36	0,24	1,74
0,32	0,10	0,06	1,72
0,23	0,15	0,11	0,25
2,31	2,92	0,83	3,14

Conf.1	
43	36
38	45

Conf.2		
43	48	28
46	48	48
49	53	50

Conf.3			
65	64	55	63
65	59	60	63
55	60	61	53
52	51	60	53

2.5 Experiment #5: Three Kinects in Front of the User

In this experiment, three Kinects are pointed in the same direction and in front of the mannequin. This experiment is complementary to the first and second experiments, and its configurations are shown in Fig. 1e. The purpose of this experiment is to determine how overlapping patterns affect the correct positioning of the mannequin. The results of this experiment are shown in Table 5.

The accuracy of the mannequin's position is lower in this experiment than in experiments #1 and #2, as shown in Fig. 2. In areas smaller than three meters, the behavior when using several Kinect agents is the same, and the accuracy is predictable. However, for distances of 4 meters the error in the positioning works in a different way. When using two Kinect agents, the results are substantially worse. But, with more sensors the results are stabilized. In parallel, the results for the erroneous pixels are also greater. In this case, we observe a difference with experiment #1. This indicates that there are problems with the sensors and, perhaps, the mannequin's positioning is not the best.

Table 5. Standard deviation of mannequin's position accuracy (in centimeters) and erroneous pixels (in thousands) in experiment #5

Conf.1		Conf.2			Conf.3			
0,06	0,07	0,04	0,09	0,05	-	0,09	0,18	-
0,08	0,08	0,05	0,12	0,28	0,06	0,27	0,06	0,09
		0,13	0,22	0,38	0,19	0,22	0,26	0,61
					3,96	3,31	5,31	5,48

Conf.1		Conf.2			Conf.3			
56	63	54	59	62	-	69	64	-
65	63	61	66	71	51	68	65	78
		66	67	66	62	67	67	76
					67	69	68	68

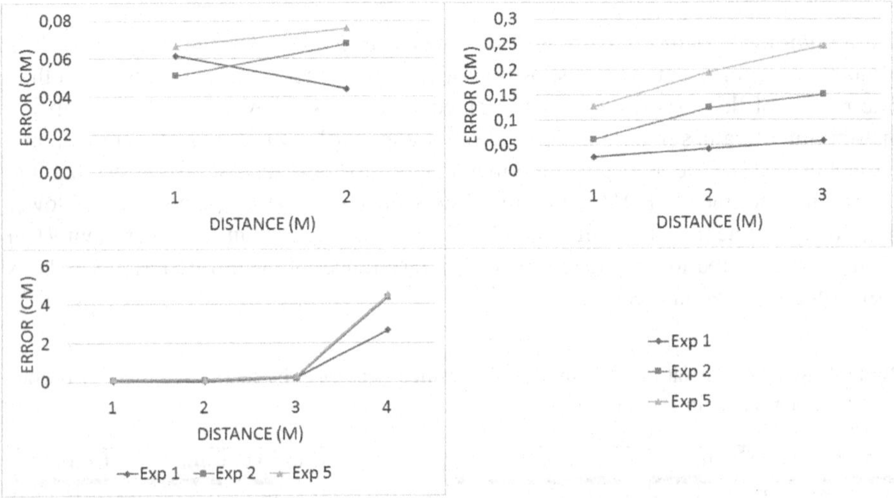

Fig. 2. Deterioration in the positioning of a user when using 1, 2 or 3 sensor agents pointing to the same direction

2.6 Experiment #6: Three Kinects Located Perpendicularly

This last experiment uses three Kinect agents located perpendicularly. The first and the third Kinect agent is one in front of the other. The configuration of this experiment is graphically represented in Fig. 1f. The purpose of this test is to determine how the interference among a large set of sensors affects the positioning of the mannequin. The results are shown in Table 6.

The results for the accuracy of the mannequin' positioning is very similar to those found for experiments #3 and #4, since this experiment is a combination of both of them. Here again, the result shows a deterioration of the accuracy (see result at the top row of experiment #4, and the right column of experiment #3). These results obtained from the combination of multiple sensors can be estimated from the base cases presented here (experiment #1). The erroneous pixels in this experiment are lower than in experiment #1. This fact emphasizes that a low erroneous frame rate does not mean that the results are really good.

Table 6. Standard deviation of mannequin's position accuracy (in centimeters) and erroneous pixels (in thousands) in experiment #6

Conf.1	
0,04	0,45
0,35	0,12

Conf.2		
0,28	0,09	0,77
0,06	0,10	0,35
0,13	0,12	0,34

Conf.3			
0,67	0,63	0,35	1,58
0,33	0,08	0,26	0,89
0,11	0,08	0,16	2,07
2,14	2,44	1,87	2,25

Conf.1	
39	30
35	36

Conf.2		
42	42	37
46	44	42
49	45	39

Conf.3			
66	65	46	47
50	52	51	47
46	52	51	42
53	51	50	41

3 Discussion and Conclusion

This work has dealt with the problem that arises when more than one Kinect sensor is used in a common recognition area. In this context, a multi-agent system approach is proposed and each Kinect has been considered an agent. This study has focused on the distance to a mannequin, and how the number and location of sensors employed affect the mannequin's position accuracy and the number of erroneous pixels. The results obtained from several experiments show the internal operation of Kinect sensors. These sensors provide better recognition data when the distance to the goal is minor than 4 meters. However, when distances are equal or greater than 4, the accuracy worsens. In this work, we have identified that the random error of depth measurement increases with increasing distance to a single Kinect sensor.

However, a combination of several Kinect sensors can improve this behavior. The mannequin's position captured by one single Kinect usually offers better results than when it is captured by more than one. Obviously, there are several situations where a single Kinect is not reliable to capture user data. This may be because the user is moving in the interaction space, or the user must be captured from different angles, or it is necessary to increase the interaction space. In these cases, it is essential to determine how to locate the cameras and how this will affect the data quality.

The use of erroneous pixels is not reliable to determine how well the obtained data is. A large number indicates that there might be problems in capturing the user, but a low number does not indicate that data has quality. However, this indicator may be used to show possible problems in the system. The results of the experiments performed should be used in the design of interaction spaces with multiple depth sensors with the aim of avoiding to make errors in the sensors' location.

Acknowledgements. This work was partially supported by Spanish Ministerio de Economía y Competitividad / FEDER under TIN2012-34003 and TIN2013-47074-C2-1-R grants, and through the FPU scholarship (FPU13/03141) from the Spanish Government.

References

1. Pavón, J., Gómez-Sanz, J.J., Fernández-Caballero, A., Valencia-Jiménez, J.J.: Development of intelligent multisensor surveillance systems with agents. Robotics and Autonomous Systems 55, 892–903 (2007)

2. Gascueña, J.M., Fernández-Caballero, A.: On the use of agent technology in intelligent, multisensory and distributed surveillance. The Knowledge Engineering Review 26, 191–208 (2011)
3. Rivas-Casado, A., Martinez-Tomás, R., Fernández-Caballero, A.: Multi-agent system for knowledge-based event recognition and composition. Expert Systems 28, 488–501 (2011)
4. Gascueña, J.M., Navarro, E., Fernández-Sotos, P., Fernández-Caballero, A., Pavón, J.: IDK and ICARO to develop multi-agent systems in support of Ambient Intelligence. Journal of Intelligent and Fuzzy Systems 28, 3–15 (2015)
5. Gonzalez-Jorge, H., Riveiro, B., Vazquez-Fernandez, E., Martínez-Sánchez, J., Arias, P.: Metrological evaluation of Microsoft Kinect and Asus Xtion sensors. Measurement: Journal of the International Measurement Confederation 46, 1800–1806 (2013)
6. Khoshelham, K., Elberink, S.O.: Accuracy and resolution of Kinect depth data for indoor mapping applications. Sensors 12, 1437–1454 (2012)
7. Oliver, M., Molina, J.P., Montero, F., González, P., Fernández-Caballero, A.: Wireless multisensory interaction in an intelligent rehabilitation environment. In: Ramos, C., Novais, P., Nihan, C.E., Corchado Rodríguez, J.M. (eds.) Advances in Intelligent Systems and Computing. AISC, vol. 291, pp. 193–200. Springer, Heidelberg (2014)
8. Chang, Y., Chen, S., Huang, J.: A Kinect-based system for physical rehabilitation: a pilot study for young adults with motor disabilities. Research in Developmental Disabilities 32, 2566–2570 (2011)
9. Oliver, M., Montero-Simarro, F., Fernández-Caballero, A., González, P., Molina, J.P.: RGB-D assistive technologies for acquired brain injury: description and assessment of user experience. Expert Systems (2015), doi:10.1111/exsy.12096
10. Mkhitaryan, A., Burschka, D.: RGB-D sensor data correction and enhancement by introduction of an additional RGB view. In: IEEE/RSJ International Conference on Intelligent Robots and Systems, pp. 1077–1083 (2013)
11. Mallick, T., Das, P.P., Majumdar, A.K.: Characterizations of noise in Kinect depth images: a review. IEEE Sensors Journal 14, 1731–1740 (2012)
12. Olesen, S.M., Lyder, S., Kraft, D., Krüger, N., Jessen, J.B.: Real-time extraction of surface patches with associated uncertainties by means of Kinect cameras. Journal of Real-Time Image Processing (2012), doi:10.1007/s11554-012-0261-x
13. Regazzoni, D., de Vecchi, G., Rizzi, C.: RGB cams vs RGB-D sensors: low cost motion capture technologies performances and limitations. Journal of Manufacturing Systems 33, 719–728 (2014)
14. Haggag, H., Hossny, M., Filippidis, D., Creighton, D., Nahavandi, S., Puri, V.: Measuring depth accuracy in RGBD cameras. In: Proceedings of the 7th International Conference on Signal Processing and Communication Systems, pp. 1–7 (2013)

A Multi-Agent System in Ambient Intelligence for the Physical Rehabilitation of Older People

Cristina Roda[1], Arturo Rodríguez[1], Víctor López-Jaquero[2], Pascual González[2], and Elena Navarro[2]

[1] Albacete Research Institute of Informatics (I3A), Albacete, Spain
{cristinarodasanchez,art.c.rodriguez}@gmail.es
[2] Computing Systems Department, University of Castilla-La Mancha, Albacete, Spain
{victor,pgonzalez,enavarro}@dsi.uclm.es

Abstract. *Ambient Intelligence* (AmI) is a very active topic of research that is gaining more and more attention because of its characteristics, transparency and intelligence. Older people is one of the collectives that can take advantage of the use of AmI systems because, thanks to these characteristics, AmI systems can focus on older adults' real needs so that they satisfy one of their main motivations to adapt technological innovations: perceived benefits. And, perhaps, everything related to healthcare and home care is perceived by them as both valuable and beneficial. In this paper, it is presented the Multi-Agent architecture (MAS) of a healthcare AmI system to treat older people' motor impairment problems by using specific devices to control the patient's movements. In this way, the natural relationship between AmI and MAS is being widely exploited. AmI proposes the development of context-aware systems that integrate different devices to recognize the context and act accordingly. Agents provide an effective way to develop such systems since agents are reactive, proactive and exhibit an intelligent and autonomous behavior. One of the main differences of our system is that it provides therapist with support to design new therapies, to adapt them to each specific person and to control their execution instead of using a fixed set of exercises.

Keywords: ambient intelligence, intelligent system, multi-agent system, healthcare AmI system, architecture, motor impairment problems, home care, telerehabilitation.

1 Introduction

Ambient Intelligence (AmI) is a very active topic of research that is gaining more and more attention because of its characteristics, *transparency* for the user and *intelligence*. As Ducatel et al. [1] state, AmI promotes the development of innovative and intelligent user interfaces "embedded in an environment that is capable of recognizing and responding to the presence of different individuals in a seamless, unobtrusive and often invisible way". This means that AmI systems become *transparent* as people do not perceive their complexity neither their presence, and are *intelligent* to react in a

J. Bajo et al. (eds.), *Trends in Prac. Appl. of Agents, Multi-Agent Sys. and Sustainability*,
Advances in Intelligent Systems and Computing 372, DOI: 10.1007/978-3-319-19629-9_13

proactive and sensitive way [2] at the same time. These two characteristics have had a great impact because it has allowed technology to be used by people who, otherwise, would have been probably computer illiterate.

Older people is one of the collectives that can take advantage of the use of AmI systems. They have been traditionally reluctant to the use of technology. For instance, a household survey [3] about ICT use carried out among 1,001 people from England and Wales in 2003 showed that the use of computer was a minority activity amongst older people, because they considered it had low relevance to their daily-life. However, AmI has become a meaningful advance in this sense, as it represents "the future vision of intelligent computing where environments support the people inhabiting them"[4], that is, AmI systems focus on older adults' real needs so that they satisfy one of their main motivations to adapt technological innovations: perceived benefits [5]. And, perhaps, everything related to healthcare is perceived by them as both valuable and beneficial.

The quality and cost of healthcare and wellbeing services is a critical issue, lately exacerbated by an increasing aging population. For this reason, the development of healthcare AmI systems are becoming a necessity that must be shortly addressed. There are already a wide diversity of these systems that are being used in different contexts. For instance, the development of the *smart homes* have allowed [6] to move away the care from hospitals to older people' home, that is, to bring the health and social care to the patient instead of bringing the patient to the health system. They have been also developed for very different purposes from training for cognitive rehabilitation [7] till physical rehabilitation [8]. This paper focuses on an AmI system for this last kind purpose in order to treat older people' motor impairment problems by using specific devices to control the patient's movements. One of the main differences of our system is that it provides therapist with support to design new therapies, to adapt them to each specific person and to control their execution instead of using a fixed set of exercises. In this paper, we present the Multi-Agent [9] architecture of this system. The rest of the paper is structured as follows. After this introduction, Section 2 presents the related work. Then, in Section 3, the architecture of the rehabilitation system is detailed. Finally, the conclusions and future work are described in Section 4.

2 Related Work

As stated by Cook et al. [10] AmI technologies are expected to be *sensitive, responsive, adaptive, transparent, ubiquitous*, and *intelligent*. Context-aware computing field mainly provides support to the first three features, while the area of ubiquitous computing is the one that facilitates transparency and ubiquity. However, *intelligence* is the most critical feature as it makes AmI systems more sensitive, responsive, adaptive, transparent and ubiquitous. The main reason is that intelligence helps in understanding user environments and, consequently, in providing adaptive assistance [11]. This explains why AmI entails contributions from different AI areas [12], such as machine learning [13], neural networks [11] and, specially, Multi-Agent System

(MAS) [9]. MAS are particularly good at modeling real-world and social systems, where problems are solved in a concurrent and cooperative way without the need of reaching optimal solutions. This is why the natural relationship between AmI and MAS is being widely exploited [14]. AmI proposes the development of context-aware systems that integrate different devices to recognize the context and act accordingly. Agents provide an effective way to develop such systems since agents are reactive, proactive and exhibit an intelligent and autonomous behavior [15]. Agents react to humans based on information obtained by sensors and their knowledge about human behaviors within agent-based AmI applications [16].

One of the approaches that combines AmI and MAS is the one proposed by Corchado et al. [17]. They propose an intelligent environment (GerAmi), which integrates MAS and other technologies, such as mobile devices, in order to facilitate the management of geriatric residences. In this sense, they assign each nurse and doctor an autonomous healthcare agent that includes relevant information about patient locations, historical data and several alarms to make a plan, so each professional can follow his agent's plan, and modify it to accommodate delays or alarms. In such a case, his associated agent makes a new planning in real time with the new information.

As stated by Isern et al. [18] there are a wide range of works that apply MAS in *healthcare*. Some of the fields of application of these agent-based approaches are *decision support systems*, which are aimed to assist the professional in the execution of healthcare processes, such as treatments or diagnostics, or *remote care*, where they are aimed to remotely monitor the status of patients allowing pervasive care. In the latter case, there are multiple proposals focusing on (remote) *home care*, e.g., to monitor and assist older people at home, identifying potential dangerous situations (AmIHomCare [19]), or to provide support in the daily activities of an older person (ROBOCARE domestic environment [20]). Another similar approach is IAServ (Intelligent Aging-in-place Home care Web Services platform) [21], which produces a personalized healthcare plan to meet the desire of patients of still living in his own house. This is done by first submitting the patient's profile to IAServ by the healthcare professional, and then this profile is converted into an ontology specification to facilitate the generation of the personalized care plan for the patient, done by an inference engine.

Other approaches are related to home care, even though they are focused on assisting professionals. For example, K4Care platform [22] provides an environment for all actors involved in the provision of home care services so they can remotely access all the knowledge they require, keep track of their current and pending activities, or request the necessary services for their patients.

Most of the approaches mentioned before [17][19][20][21] are specifically oriented to *older people* as they represent a susceptible population group to be assisted regarding healthcare and home care. So much so that there are a wide variety of Ambient-Assisted Living (AAL) tools for older adults based on the AmI paradigm, as Rashidi and Mihailidis stated in [23]. These authors distinguish several AAL application areas, but we are especially interested in one in particular: *Therapy*, namely in telerehabilitation systems as it is the main object of our research.

Regarding to *rehabilitation* area, we can find some works that propose intelligent robotic systems to assist the physical rehabilitation process of the patient, e.g., for knee rehabilitation [8] or for lower limb rehabilitation [7] that, unlike the first one, it uses a MAS to detect bioelectric and physical signals through a sensor network located in the patient's body in order to determine his movement intention and assist him in doing such a movement. Another different proposal, called OntoRis [24], offers an ontology-based rehabilitation service that the patient can use to acquire comprehensive information about his prescribed rehabilitation treatment, or it can simply serve as an interactive learning platform for people interested in this particular medical field. Finally, Abreu et al. [25] focus their attention on cognitive rehabilitation, namely on using 3D games for neuropsychiatric disorders rehabilitation. These authors propose a MAS for automatic control while the patient is playing a 3D game in order to reduce human intervention needed to manage the execution of software processes.

In this sense, our proposal has some similarities with respect to the one presented by Abreu et al. as we also propose a *MAS* that is able to control the performance of all the tasks that a patient is doing during a rehabilitation therapy. The main difference here is that our approach is centered on *physical rehabilitation*, instead of cognitive rehabilitation. However, unlike the works mentioned before [8][7] about physical rehabilitation where an intelligent controller is used to manage the robot behavior, our proposal is focused on the performance of the tasks prescribed in a particular rehabilitation therapy.

Moreover, our proposal is centered on physical rehabilitation for *older people* as we have noticed that this collective has specific difficulties while doing rehabilitation exercises mainly because of physical (and/or psychical) problems associated with aging.

Furthermore, our system takes the advantage of using a MAS combined with *AmI*, as some of the works mentioned before. Therefore, the use of the AmI paradigm makes sense here when talking about older people given that AmI provides transparent and intelligent mechanisms to interact with any type of software. The benefits of using AmI for older people are clearly stated before, allowing an older adult to interact in a transparent, simple and easy way with our rehabilitation system. This kind of intelligent system may avoid possible conflicts that arise when older people interact with software systems in a classical way as most of them are not familiarized with the use of technology at all.

Thus, this paper presents an extension and enrichment of the system proposed in a previous work [26] in order to reach the development of a complete system able to create, perform and monitor therapies for physical rehabilitation of older people diseases. Our proposal provides a tool that can be used by experts in the field of rehabilitation to define a set of customized therapies and the rules that determine the behavior of the system at runtime. In this way, activities can be adapted to older people while performing a particular therapy. The creation of therapies is driven by a metamodel that defines the Domain Specific Language. Namely, in rehabilitation domain, a *Therapy* is composed by Activities, an *Activity* is composed by Tasks, and a *Task* is a set of *Steps* which can be Gestures or Postures that the older person has to perform. Relationships between elements of the same hierarchical level can be established in order to define a sequence of Therapies, Activities or Tasks, respectively. Therefore, the therapy model can be described as a composed diagram state.

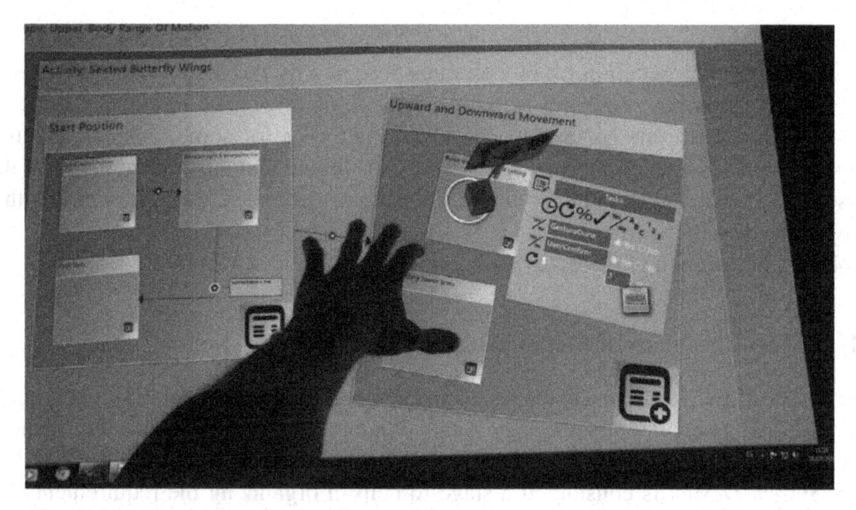

Fig. 1. Snapshot of the user interface to define a therapy, using Microsoft PixelSense

The therapist can easily instantiate the metamodel by using a user interface developed for Microsoft PixelSense (see Fig 1) and specify the movements that the older person has to perform by using Microsoft Kinect. The resulting model can be interpreted by the system which automatically generates therapies that are performed by older people with motor impairment problems. These therapies are also supervised and monitored by the system using MS Kinect and other kind of sensors that provide information about physical conditions of the older person in a transparent way. Furthermore, in rehabilitation context, an older person only has to move according to the specified therapy without direct interaction with any devices, taking advantage of the AmI paradigm. Moreover, the therapist defines a set of Fuzzy Inference Systems (FISs) that allows our system to adapt the therapies and to decide the performance order of the tasks, activities and therapies at runtime. In order to support these features, a MAS has been integrated into our system, which will be described in the next section.

3 A MAS Architecture for Older People Rehabilitation in AmI

In this section, a general overview of our MAS designed to improve the supervision and execution of therapies for older people is discussed. It makes use of AmI to enrich the environment provided to older people and improve the decision making activities derived from the requirement to provide therapies adapted to the ambient situation and the person using the system.

Our MAS is focused on the rehabilitation of older people diseases related to physical problems. In this rehabilitation process, a therapist prescribes a treatment comprising several therapies designed for the rehabilitation of specific complains. Therapies are organized according to activities, and those activities consists of some tasks. Finally, these tasks are carried out by executing some steps (in our case, these steps can be either gestures or postures). Thus, this MAS is designed to achieve several goals: (1) it supports the execution of the therapies designed by the therapist for older people, (2) it carries out sensoring activities to monitor how well an activity is being done. Furthermore, this MAS also provides monitoring for some ambient features, namely the oxygen level, the stress, the beats per minute and the current mood of the person. All these features are captured by means of dedicated hardware devices that provide the data to the MAS.

Next, a description of the overview of the design of our MAS is included.

3.1 MAS Development Methodology

The design of our MAS was made by using Prometheus Design Tool [27], the tool for supporting Prometheus methodology for intelligent agent development. Prometheus comprises three main stages: System Design, Architectural Design and Detailed Design. *System Design* is considered a stage to help in organizing the requirements for the MAS by expressing them into a comprehensive graphical notation. This phase is made based on the principles of the goal-orientation approach. In *Architectural Design* stage the composition of the MAS is designed. This stage includes deciding what agents the MAS will have, the messages they will exchange or the capabilities they will exhibit. Finally, *Detailed Design* stage is related to the internal design of agents. As these stages are made for all the artefacts involved in the MAS, the System Overview diagram is composed. This diagram is a great tool for representing the general architecture of a MAS system (see Fig 2). This graphical notation includes the agents involved (in light brown color), the protocols including the messages they exchange (in magenta color), the actions each agent is responsible for (in light green color) and lastly, the percepts that represent the external information arriving to the MAS (in light red color) from the ambient.

3.2 Managing Ambient Information

In Fig 2, the overview of the design of our MAS system is depicted. To manage the external information arriving from the ambient to the MAS, dedicated agents are used. We decided to use one agent for each type of external ambient information (percept), because it is easier to extend its capabilities. This is an important feature, since the availability of sensors for assorted sources of information advances very quickly. Thus, our MAS must be ready to easily consider extra valuable information from the ambient to improve the available information used in the decision making process. Therefore, we have an agent for each type of information arriving to the MAS from the ambient. The data sources we are using are oxygen level, posture, gesture, stress, BPM and mood. *AgentOxygenLevelCapture, AgentPostureCapture, AgentGestureCapture, AgentStressCapture, AgentBPMCapture* and *AgentMoodCapture* are the agents responsible for capturing the information from the ambient, respectively.

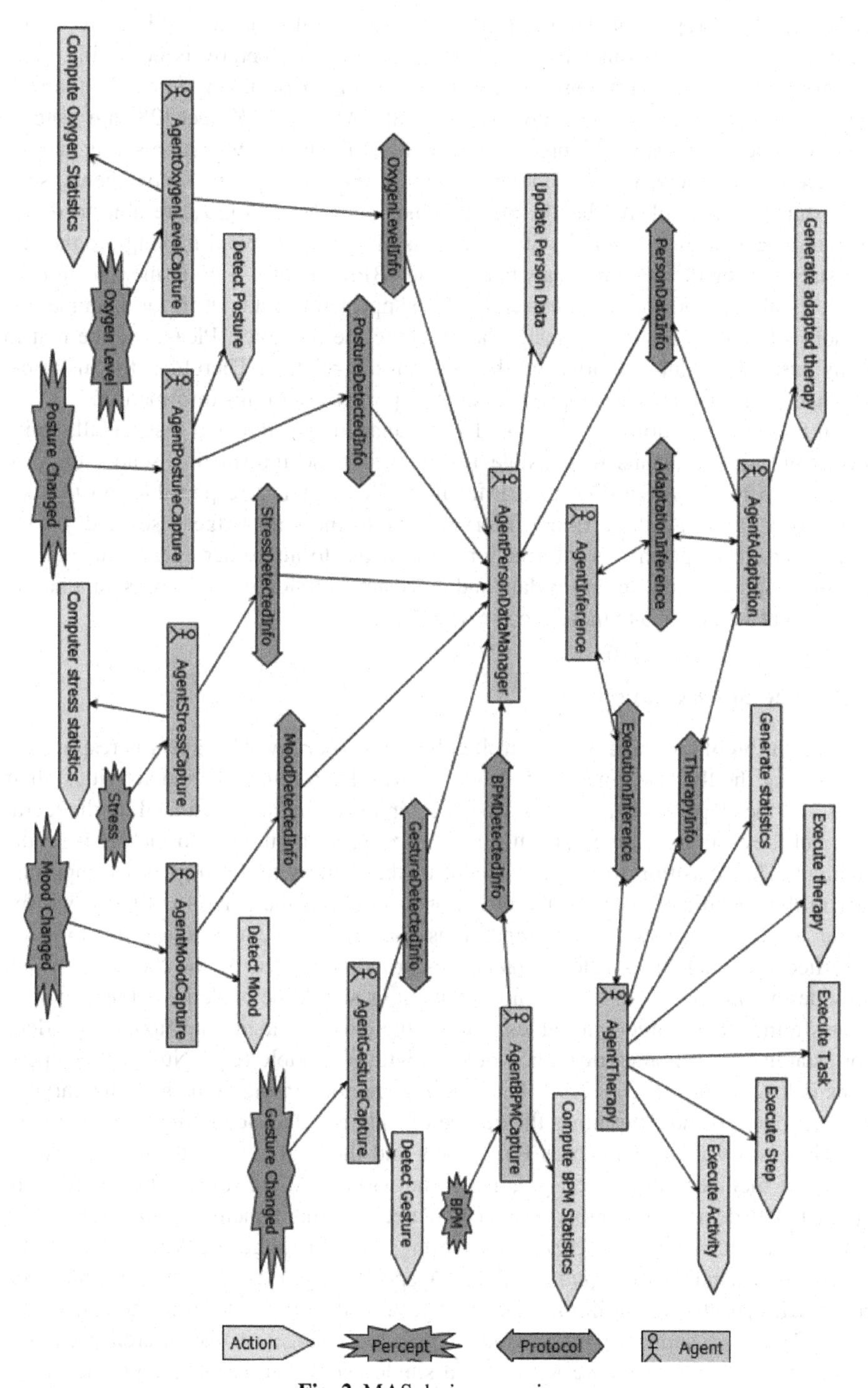

Fig. 2. MAS design overview

Oxygen level represents the saturation of oxygen in the blood, and it is a good indicator to help to find out when the person making the therapy is becoming tired. Posture represents the different postures that the person is making to do the therapy, e.g., raising the arms or the elbow. Devices like Microsoft Kinect [28] are used to capture these postures, which provide information about how well a person is making each step of the therapy. Some therapies require more fine-grained movements, such as gestures made with the hands. Devices like Leap Motion [29] are able to detect gestures made with both hands. Stress level is also used to avoid making the therapy too stressful for the person. Beats per minute (BPM) is also a good measure for fatigue. Finally, mood change is currently also supported to find out if, for example, the person is happy, excited or pained when making the therapies. Please, notice that in many cases the values reported by the ambient are relative. Therefore, the historical data and some statistics are required to make a proper use of this information.

All these information must be stored and managed. To store and manage all the information, all the agents responsible for capturing the information send their processed data to the AgentPersonDataManager. This agent is responsible, on the one hand, to integrate all the arriving information into the knowledge bases and, on the other hand, to dispatch this information on demand to any other agent that requires this information. Thus, an up-to-date and consistent version of the data is supplied to every agent so as to make more precise decisions.

3.3 Therapy Execution

The execution of the therapy is controlled by *AgentTherapy*. This agent is responsible for running the therapy. Therefore, it will control the execution of the activities, their tasks and the steps these tasks prescribe. Therapies are initially designed by the therapist, but they can be adapted at run-time according to the input information from the ambient. The transition from one element of the therapy to the next is an important fact in the specification of the therapy. For example, if the person is tired it will be better to reduce the number of repetitions for each element. These transitions are specified by using fuzzy rules. Fuzzy rules are also created by the therapist, which provides a natural way to express how transitions should be made by using linguistic values rather than numerical values. Unlike the classical logic, the fuzzy logic does not establish rigid boundaries when dealing with a variable (e.g., Number of repetitions is low when it is 3 or less), but it defines regions in the domain of the variable that can overlap and determines the degree of truth of a fact according to the real value. Those regions are known as Fuzzy Sets and the degree of truth of a fact is the degree of membership of a crisp value of the variable to the fuzzy set referred to in the fact. The degree of membership is calculated by using a membership function that defines that fuzzy set. The fuzzy sets are tagged with Linguistic Values. All fuzzy inference is handled by *AgentInference*. A specific agent is used for the inference engine aiming at keeping the inference as separated as possible from the rest of the agent. Thus, we support the isolation of possible changes in the inference engine. Furthermore, in this way, we have also a single agent that access the profile of the person. This is also interesting from the point of view of privacy, because by securing a single agent, the privacy of the data of the persons is improved, thus reducing the development effort for privacy issues. Privacy has been proven an issue for any kind of medical related software, since most people find their medical information especially private.

3.4 Logging Therapy Statistics

Another action carried out by *AgentTherapy* is generating statistics. For therapist to follow the progress of their patients, they need to get feedback about the execution of the therapies they design [30]. Aiming to provide such feedback to the therapist, *AgentTherapy* logs every action the patient does during the execution of the therapy, logging every activity, task and step the patient took during the therapy. These statistics include currently completion time, repetitions, postures and gestures identified, tries before a posture or a gesture is made ok, and fatigue. With these statistics, the therapist can follow how the patients progress and they will provide valuable feedback to design other therapies or adapt the current ones.

3.5 Adapting Therapies

Although the therapists have feedback from the logging to try to design the therapy, it is impossible to foresee all the ambient factors that can have an influence during the execution of the therapy. Therefore, the *transitions* from one element of the therapy to another is made in terms of fuzzy rules. These fuzzy rules support the therapist in designing under what circumstances the transition should be to one element or another, thus supporting and adaptive workflow of the therapy execution. These fuzzy rules are evaluated by the *AgentInference* agent.

On the other hand, therapy execution can also be adapted in another manner. In this case, *AgentAdaptation* makes use of the information gathered by *AgentPersonDataManager* from the information captured through the sensors and the therapy execution logs to adapt the *fuzzy sets* to reflect the issues detected. For instance, if *TherapyAgent* reports that the user has made too many repetitions of a gesture or a posture to do it right, *AdaptationAgent* decides to reduce the precision required to consider the posture or gesture as right by adapting the corresponding fuzzy sets. This loss of precision can be because of fatigue, therefore *AdaptationAgent*, by checking the evolution of the data captured by the sensors, can decide to recommend a break in the therapy execution to the patient. Thus, we provide a therapy execution environment that can cope with a vast number of ambient situations that can change the initially predicted therapy execution defined by the therapist.

4 Conclusions and Future Work

In this paper, a new system for monitoring the execution of therapies in an Ambient Intelligence Environment by using a multi agent system is presented. This system allows the control of some specific physical therapies for the rehabilitation of older people diseases. These therapies are designed by therapists from scratch by defining, not only the elementary activities, but also the rules that determine the behavior of the system. Their design is driven by a metamodel that defines the Domain Specific Language of the rehabilitation environment and includes a set of Fuzzy Inference Systems that allows our system to adapt the therapies and to decide the performance order of the tasks, activities and steps at runtime.

In order to support these features, a MAS has been designed. It makes use of AmI to enrich the environment provided to older people and improve the decision making activities derived from the requirement to provide therapies adapted to the ambient situation and the person using the system. Therefore, the MAS provides monitoring for some ambient features, namely the oxygen level, the stress, the beats per minute and the current mood of the person. All these features are captured by means of dedicated hardware that provide the data to the MAS. In addition, our system provides to the therapists some feedback about the execution statistics that allows monitoring the execution and evolution of the older people that should carry out these therapies.

Our next step is the evaluation of our system in a real environment. To carry out this evaluation we will have the support of some associations that assist older people and people affected by acquired brain injury.

Acknowledgements. This work was partially supported by the Spanish Ministry of Economy and Competitiveness and by the FEDER funds of the EU under the project grant insPIre (TIN2012-34003). It has also been funded by the Spanish Ministry of Education, Culture and Sport thanks to the FPU scholarship (FPU12/04962).

References

1. Ducatel, K., Bogdanowicz, M., Scapolo, F., Leijten, J., Burgelman, J.C.: ISTAG: Scenarios for ambient intelligence in 2010. Technical Report, ISTAG (2010)
2. Augusto, J.: Ambient intelligence: the confluence of pervasive computing and artificial intelligence. In: Intell. Comput. Everywhere, pp. 213–234 (2007)
3. Selwyn, N., Gorard, S., Furlong, J., Madden, L.: Older adults' use of information and communications technology in everyday life (2003)
4. Acampora, G., Cook, D.J., Rashidi, P., Vasilakos, A.V.: A survey on ambient intelligence in healthcare. Proc. IEEE. 101, 2470–2494 (2013)
5. Melenhorst, A.-S., Rogers, W.A., Bouwhuis, D.G.: Older adults' motivated choice for technological innovation: evidence for benefit-driven selectivity. Psychol. Aging 21, 190–195 (2006)
6. Augusto, J., Mccullagh, P.: Ambient Intelligence: Concepts and applications (2007)
7. Chen, J., Zhang, X., Li, R.: A novel design approach for lower limb rehabilitation training robot. In: 2013 IEEE International Conference on Automation Science and Engineering (CASE), pp. 554–557 (2013)
8. Akdoğan, E., TaçgIn, E., Adli, M.A.: Knee rehabilitation using an intelligent robotic system. J. Intell. Manuf. 20, 195–202 (2009)
9. Gascueña, J.M., Navarro, E., Fernández-Caballero, A.: Model-driven engineering techniques for the development of multi-agent systems. Eng. Appl. Artif. Intell. 25, 159–173 (2012)

10. Cook, D.J., Augusto, J.C., Jakkula, V.R.: Ambient intelligence: Technologies, applications, and opportunities. Pervasive Mob. Comput. 5, 277–298 (2009)
11. Chernbumroong, S., Cang, S., Atkins, A., Yu, H.: Elderly activities recognition and classification for applications in assisted living. Expert Syst. Appl. 40, 1662–1674 (2013)
12. Ramos, C., Augusto, J.C., Shapiro, D.: Ambient Intelligence—the Next Step for Artificial Intelligence. IEEE Intell. Syst. 23, 15–18 (2008)
13. Aztiria, A., Izaguirre, A., Augusto, J.C.: Learning patterns in ambient intelligence environments: a survey. Artif. Intell. Rev. 34, 35–51 (2010)
14. Gascueña, J.M., Navarro, E., Fernández-Sotos, P., Fernández-Caballero, A., Pavón, J.: IDK and ICARO to develop multi-agent systems in support of Ambient Intelligence. J. Intell. Fuzzy Syst. (2014)
15. Ayala, I., Amor, M., Fuentes, L.: A model driven engineering process of platform neutral agents for ambient intelligence devices. Auton. Agent. Multi. Agent. Syst. 28, 214–255 (2013)
16. Bosse, T., Both, F., Gerritsen, C., Hoogendoorn, M., Treur, J.: Methods for model-based reasoning within agent-based Ambient Intelligence applications. Knowledge-Based Syst. 27, 190–210 (2012)
17. Corchado, J.M., Bajo, J., Abraham, A.: GerAmi: Improving Healthcare Delivery in Geriatric Residences. Intell. Syst. 23, 19–25 (2008)
18. Isern, D., Sánchez, D., Moreno, A.: Agents applied in health care: A review. Int. J. Med. Inform. 79, 145–166 (2010)
19. Mocanu, S., Mocanu, I., Anton, S., Munteanu, C.: AmIHomCare: a complex ambient intelligent system for home medical assistance. In: 10th International Conference on Applied Computer Science, pp. 181–186 (2011)
20. Cesta, A., Cortellessa, G., Rasconi, R., Pecora, F., Scopelliti, M., Tiberio, L.: Monitoring elderly people with the ROBOCARE domestic environment: Interaction synthesis and user evaluation. Comput. Intell. 27, 60–82 (2011)
21. Su, C.J., Chiang, C.Y.: IAServ: An intelligent home care web services platform in a cloud for aging-in-place. Int. J. Environ. Res. Public Health 10, 6106–6130 (2013)
22. Isern, D., Moreno, A., Sánchez, D., Hajnal, Á., Pedone, G., Varga, L.Z.: Agent-based execution of personalised home care treatments. Appl. Intell. 34, 155–180 (2011)
23. Rashidi, P., Mihailidis, A.: A survey on ambient-assisted living tools for older adults. IEEE J. Biomed. Heal. Informatics 17, 579–590 (2013)
24. Su, C.J., Peng, C.W.: Multi-agent ontology-based Web 2. 0 platform for medical rehabilitation. Expert Syst. Appl. 39, 10311–10323 (2012)
25. de Abreu, P.F., Werneck, V.M.B., de Costa, R.M.E.M., De Carvalho, L.A.V.: Employing multi-agents in 3-D game for cognitive stimulation. In: Proceedings - 2011 13th Symposium on Virtual Reality, SVR 2011, pp. 73–78 (2011)
26. Rodríguez, A.C., Roda, C., Montero, F., González, P., Navarro, E.: A Collaborative System for Designing Tele-Therapies. In: Pecchia, L., Chen, L.L., Nugent, C., Bravo, J. (eds.) IWAAL 2014. LNCS, vol. 8868, pp. 377–385. Springer, Heidelberg (2014)
27. RMIT Intelligent Agents Group: Prometheus/PDT, https://sites.google.com/site/rmitagents/software/prometheusPDT
28. Microsoft: Kinect for Windows
29. Leap Motion Inc: Leap Motion
30. Navarro, E., López-Jaquero, V., Montero, F.: HABITAT: A Web Supported Treatment for Acquired Brain Injured. In: Eighth IEEE International Conference on Advanced Learning Technologies, ICALT 2008, pp. 464–466. IEEE (2008)

A Distributed Architecture for Multimodal Emotion Identification

Marina V. Sokolova[1], Antonio Fernández-Caballero[1], María T. López[1], Arturo Martínez-Rodrigo[2], Roberto Zangróniz[2], and José Manuel Pastor[2]

[1] Universidad de Castilla-La Mancha, Instituto de Investigación en Informática de Albacete, 02071-Albacete, Spain
[2] Universidad de Castilla-La Mancha, Instituto de Tecnologías Audiovisuales, 16071-Cuenca, Spain

Abstract. This paper introduces a distributed multiagent system architecture for multimodal emotion identification, which is based on simultaneous analysis of physiological parameters from wearable devices, human behaviors and activities, and facial micro-expressions. Wearable devices are equipped with electrodermal activity, electrocardiogram, heart rate, and skin temperature sensor agents. Facial expressions are monitored by a vision agent installed at the height of the human's head. Also, the activity of the user is monitored by a second vision agent mounted overhead. The emotion is refined as a cooperative decision taken at a central agent node denominated "Central Emotion Detection Node" from the local decision offered by the three agent nodes called "Face Expression Analysis Node", "Behavior Analysis Node" and "Physiological Data Analysis Node". This way, the emotion identification results are outperformed through an intelligent fuzzy-based decision making technique.

Keywords: Distributed multiagent system, multimodal architecture, emotion identification.

1 Introduction

Nowadays, people interact constantly with information technology (IT) devices, both at their professional activity and at leisure time. The qualitative part of this interaction takes a special significance, just as it is directly related to the emotional state of the user. In this sense, the emotional comfort, as well as emotion detection and recognition, is becoming an essential part of any technological or computerized application (e.g. [1], [2]). Emotion detection can be provided by capture and analysis of implicit information about the user, such as voice, facial changes, alterations of body movements and gestures, among others (e.g. [3], [4]).

Recognition of affective behavior is a part of affective interaction, in which a computer aims at approaching an emotional state, mood, attitude or personal preference of a human user. When adding the ability of emotional intelligence to computers, this makes them more coherent to humans who interact with them. Emotion recognition and management is a core part of affective computing, a quickly developing branch of science (e.g. [5], [6]), which shares spheres of research and integrates methods and approaches

J. Bajo et al. (eds.), *Trends in Prac. Appl. of Agents, Multi-Agent Sys. and Sustainability*,
Advances in Intelligent Systems and Computing 372, DOI: 10.1007/978-3-319-19629-9_14

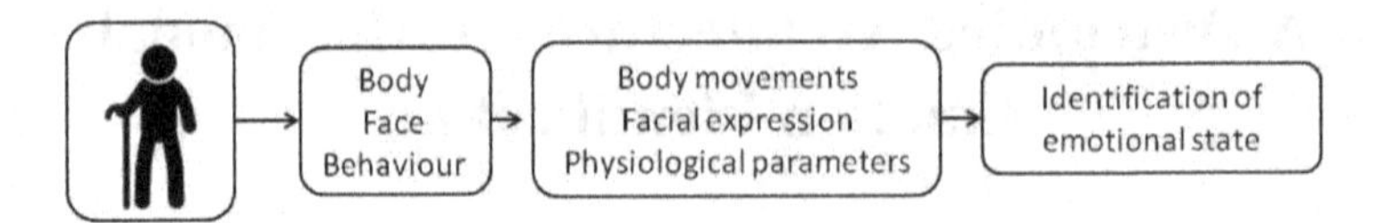

Fig. 1. System of the distributed emotion detection architecture

from computer science, psychology, physiology, and so on. Many publications in this area coincide that one of the important findings is that comfort and well-being of a user is reached by providing affective modulation.

There are different types of emotion recognition systems, starting with unimodal ones, such as face or voice expression recognition systems [7], and systems based on processing physiological parameters. Among the latter, in [8] electrocardiogram, skin temperature variation and electrodermal activity are registered and analyzed, and in [9] electromyogram is used to identify emotions. Now, multimodal systems combine the registration of various types of parameters [10]. Generally, multimodal systems show better performance results as they are more error-prone, and they provide multiple independent sources of information. Also, in case of error, there should be disagreement between the outcomes, though, on the contrary case, each module proves and confirms the common result (e.g. [11], [12]).

This paper presents a distributed multiagent system architecture (MAS) for multimodal emotion detection, encompassing non-invasive emotion detection through cameras and data collected by wearable sensors [13]. Wearable devices are equipped with electrodermal activity, electrocardiogram, heart rate, and skin temperature sensor agents. Facial expressions are monitored by a vision agent installed at the height of the human's head. Also, the activity of the user is monitored by a second vision agent mounted overhead. So, emotion detection is realized as shown in Fig. 1. Here, emotional changes experienced by a person are reflected in his/her face, as well as they can also alter activities as well as provoke other internal bodily changes (see, [14], [4]) as, for instance, an increase/discrease of the heart rate. Wearable sensors obtain data on physiological indicators and also register changes in parameters.

The emotion is refined as a cooperative decision taken at a central agent node denominated "Central Emotion Detection Node" from the local decision offered by the three agent nodes called "Face Expression Analysis Node", "Behaviour Analysis Node" and "Physiological Data Analysis Node". The final emotional state is determined as a composite decision between the outputs generated by the three independent sources. In this manner, an emotion is represented as a pattern which includes spatial, temporal, and physiological indicators.

2 Emotion Identification Architecture

Our proposal of an emotion identification distributed architecture (EIDA) is modeled as a distributed agent network which incorporates a set of nodes. Each of the agent node performs its own autonomous functions. Initial information is gathered on the logical hardware level, which is in charge of capturing data from independent sensors

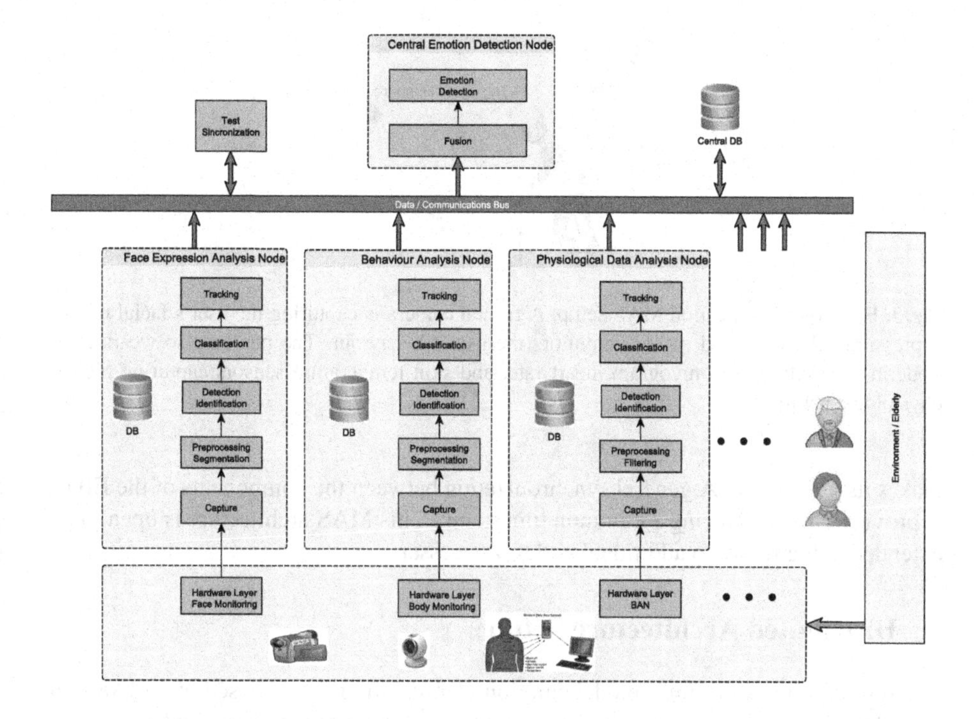

Fig. 2. Layout of the emotion identification distributed architecture

(video cameras and physiological wearable sensors, modeled as intelligent and physical agents) and transferring them to the proper nodes. Fig. 2 shows the structure of the EIDA and its components. The architectures includes the following agent nodes:

- "Hardware Level" which consists of independent sensor agents. They monitor people's behaviours, capture their facial expressions, and collect physiological data from wearable sensors.
- "Face Expression Analysis Node" which is in charge of constant monitoring and identification of human's facial expressions.
- "Behaviour Analysis Node" which executes a set of data processing methods aiming at the identification, classification, and tracking of human behaviours.
- "Central Emotion Detection Node" which fuses the outputs of the analysis nodes received through the "Data/Communication Bus", and takes the decision on which emotion is perceived from a person.

Each node is composed of a pre-established set of levels. The levels show a natural work-flow for image/signal processing, starting with data acquisition and preprocessing, followed by identification and detection, and concluding with classification and tracking. Additionally, each node performs an initial classification of emotions based on its autonomous reasoning. In case it is not possible to identify an emotion or there is a disagreement in classification between nodes, the "Central Emotion Detection Node"

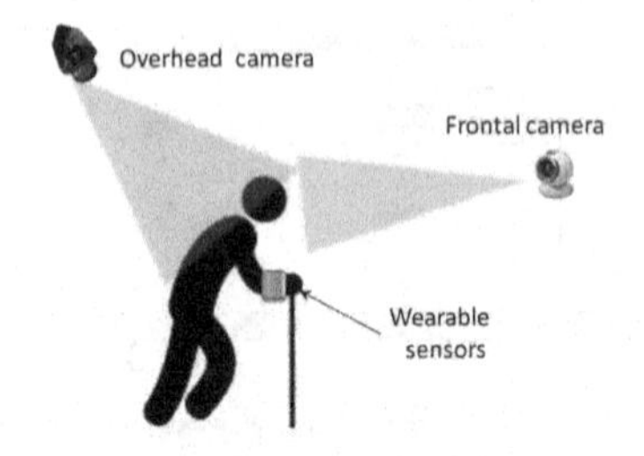

Fig. 3. Emotion identification MAS setup. A frontal camera is capturing the user's facial micro-expressions. An overhead camera monitors the user's behaviour. The person also wears electrodermal activity, electromyogram, heart rate, and skin temperature sensors capturing his/her physiologic parameters.

makes its judgement. A general synchronization between the components of the EIDA is provided by introducing a common timestamp. This MAS architecture is open to be extended in the future to additional nodes if necessary.

3 Distributed Architecture Setup

The setup proposed for multimodal emotion identification is composed of a MAS that provides constant monitoring of physical activity, facial expressions, and control of physiologic parameters (see Fig. 3). The experimentation area (testing environment) is equipped with an overhead and a frontal camera. The first camera is mounted to cover the movements of a person, whilst the second camera captures his/her facial expression. The user is wearing sensors which transmit data in order to detect changes in the main physiologic indicators related to emotions.

3.1 Facial Emotion Recognition and Behaviour Analysis

The human face is the best source that reflects multiple emotions and brings almost the necessary information for affect recognition. As it has been proved in many studies, the visual channel is dominant in affective stimuli perception as well as for emotional state "reading" through visually detectable indicators (e.g. [15], [16]). Thus, the interpretation of facial emotions and their correct classification accomplishes a significant part of affective status identification [17].

Our proposal stars from the Facial Action Coding System (FACS) which offers a detailed decomposition of facial muscle movements and classifies them into so-called action units [18]. Action units are related to the movements of specific sets of facial muscles that produce the expression. The emotion is recognized though using support vector machines (SVM) to train a classifier. The classifier is trained and tested on the standard databases JAFFE[1], MMI[2], and Cohn-Kanade[3], and show high performance

[1] `http://www.kasrl.org/jaffe.html`

[2] `http://mmifacedb.eu/`

[3] `http://www.pitt.edu/~emotion/ck-spread.htm`

results. The facial emotion detection node executes in the following way. A person is continuously monitored while carrying out his/her daily activity. Micro-movements of facial muscles are captured, evaluated, and classified.

Now, the behaviour is also closely related to emotions as it has been previously reported (see, for instance, [19]). Indeed, different emotional patterns provoke arousal or relaxation of specified groups of muscles, as well as changes in velocity, acceleration and trajectory of movements. Depending on the emotion that is being experienced, a person becomes excited or ceases activities, and his/her posture also serves as an indicator of his/her mood. Behavior identification and tracking is gotten through constant monitoring. The obtained data are preprocessed with image preprocessing algorithms within the OpenCV library[4] for real-time computer vision applications.

3.2 Physiological Sensor Data Interpretation

The data set consists of measurements of four physiological signals, which include electro-dermal activity (EDA), superficial electromyogram (EMG), heart rate variability (HRV) and skin temperature (SKT). These markers are chosen due to their suitability reporting the arousal level on the user.

1. EDA is a measure of the skin conductivity, and its value is related with the increment in the sweat glands' activity, as the skin is better able to conduct electricity. The sweat secretion is also controlled by the sympathetic nervous system which reacts against any stress situation, pain or mental illness.
2. EMG represents the electrical activation of the muscles, which is controlled by the nervous system. It has been reported that motor activation may be helpful in measuring the emotions, since evidence indicates that voluntary and involuntary movements are modulated by the emotional context [20], as well as the innate disposition of the individual to move more rapidly in stress contexts, making EMG signals helpful to reveal the excitation degree of the person [21].
3. HRV shows the irregularities produced on the heart, by computing the temporal distance between consecutive heart cycles. Given that heart activity is controlled by the autonomous nervous system, HRV measurements are appropriate to quantify the arousal level on the individual. Indeed, HRV has been identified as a direct link over the autonomic nervous system, being affected by the sympathetic nervous, such that an increasing HRV can be observed when the person is under stress stimuli.
4. SKT: This parameter is another indicator of arousal degree. Its decreasing and increasing has been shown to be related to specific emotions [22]. Indeed, when the user is under stress, body muscles are tensioned, provoking the blood vessels to contract and, consequently, there is a decreasing of the skin temperature.

Data is obtained by the "Wearable Physiological Data Acquisition System" (WPDAS). In this regard, WPDAS is controlled by an ultra-low-power, 32-bit ARM Cortex-M3 microcontroller (UC). The device architecture is chosen after taking into consideration the low power consumptions and its scalability to another more powerful UC's.

[4] `http://opencv.org/`

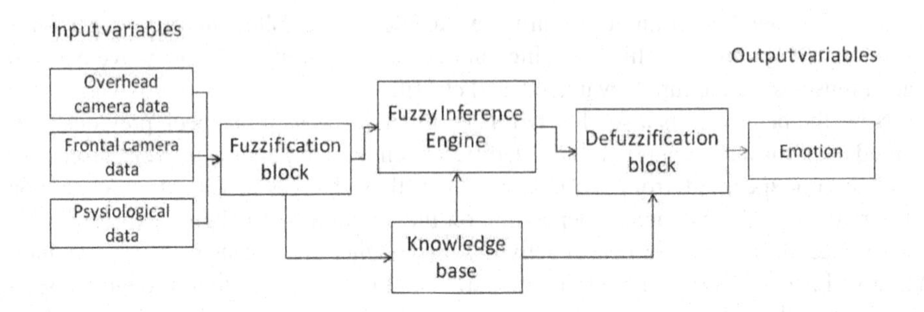

Fig. 4. Fuzzy inference system for emotion detection

Furthermore, different signal acquisition systems are used to adequate the physiological variables before being sampled by the UC.

3.3 Fusion for Emotion Recognition

Fuzzy logic is one of the artificial intelligence are to solve decision making under highly uncertain conditions. It enables implementation of "human-like" reasoning, which is optimal for multiagent systems such as the distributed architecture proposed in this paper. Thus, a fuzzy-based fusion mechanism is proposed to enhance the overall system performance. Fuzzy logic eases the problem of emotional state identification, on the one hand, and facilitates recognition of emotional patterns, on the other hand. The proposed fuzzy inference system is presented in Fig. 4.

The input variables, which include crisp data from the three independent sources (overhead and frontal cameras and from the WPDAS) are transformed into linguistic variables within the "Fuzzification block". On the next step, the "Fuzzy inference engine" simulates the reasoning process using fuzzy "IF-THEN" rules and data from the "Knowledge base". The fuzzy rules are generated based on the expert's knowledge and using linguistic variables. Lastly, the fuzzy set obtains the crisp values corresponding to the output crisp variable "Emotion". This variable offers an integer value from 1 to 7, which correspond to one of the basic emotions described by Ekman.

To fuse information which is received on the input of the "Fuzzification block", for the behaviour and physiological data analysis respectively the following linguistic variables are stated:

- "The amount of moving", which has the terms *Active*, *Reserved*, and *Low*.
- "The immobility time", with the terms *High*, *Medium*, and *Low*.
- "EDA" with the terms *High*, *Medium*, and *Low*.
- "EMG" represented with the terms *Tensed* and *Relaxed*.
- "HRV", which includes the terms *Regular* and *Irregular*.
- "SKT", with the terms *High*, *Normal*, and *Low*.

The Face Expression Analysis Node outputs a variable which contains information on emotional state, and a probability of the given emotion. These data is also given on the input of the 'Fuzzification block".

4 Conclusions

A multimodal distributed MAS architecture for emotion identification has been described. The architecture is based on simultaneous and autonomous analysis of independently processed physiological parameters and visual data. The visual agents capture facial expressions and activities/behaviours. The physiologic signals include electrodermal activity, superficial electromyogram, heart rate variability and skin temperature. Finally, a fuzzy inference system at central agent node "Central Emotion Detection Node" identifies the emotional pattern of a person, fusing all available data coming from the agent nodes "Face Expression Analysis Node", "Behaviour Analysis Node" and "Physiological Data Analysis Node".

Acknowledgements. This work was partially supported by Spanish Ministerio de Economía y Competitividad / FEDER under TIN2013-47074-C2-1-R grant.

References

1. Fernández-Caballero, A., Latorre, J.M., Pastor, J.M., Fernández-Sotos, A.: Improvement of the elderly quality of life and care through smart emotion regulation. In: Pecchia, L., Chen, L.L., Nugent, C., Bravo, J. (eds.) IWAAL 2014. LNCS, vol. 8868, pp. 348–355. Springer, Heidelberg (2014)
2. Castillo, J.C., Fernández-Caballero, A., Castro-González, Á., Salichs, M.A., López, M.T.: A framework for recognizing and regulating emotions in the elderly. In: Pecchia, L., Chen, L.L., Nugent, C., Bravo, J. (eds.) IWAAL 2014. LNCS, vol. 8868, pp. 320–327. Springer, Heidelberg (2014)
3. Darwin, C.R.: The Expression of the Emotions in Man and Animals. London, John Murray (1872)
4. Karg, M., Samadani, A., Gorbet, R., Kuhnlenz, K., Hoey, J., Kulic, D.: Body movements for affective expression: a survey of automatic recognition and generation. IEEE Transactions on Affective Computing 4, 341–359 (2013)
5. Picard, R.W.: Affective computing: challenges. International Journal of Human-Computer Studies 59, 55–64 (2003)
6. Tao, J., Tan, T.: Affective computing: a review. In: Tao, J., Tan, T., Picard, R.W. (eds.) ACII 2005. LNCS, vol. 3784, pp. 981–995. Springer, Heidelberg (2005)
7. Petrushin, V.A.: Emotion recognition in speech signal: experimental study, development, and application. In: 6th International Conference of Spoken Language Processing, pp. 222–225 (2000)
8. Kim, K.H., Bang, S.W., Kim, S.R.: Emotion recognition system using short-term monitoring of physiological signals. Medical and Biological Engineering and Computing 42, 419–427 (2004)
9. Murugappan, M.: Electromyogram signal based human emotion classification using KNN and LDA. In: 2011 IEEE International Conference on System Engineering and Technology, pp. 106–110 (2011)
10. Bänziger, T., Grandjean, D., Scherer, K.R.: Emotion recognition from expressions in face, voice, and body: the multimodal emotion recognition test (MERT). Emotion 9, 691–704 (2009)
11. Picard, R.W.: Emotion research by the people, for the people. Emotion Review 2, 250–254 (2010)

12. Kukolja, D., Popovic, S., Horvat, M., Kovac, B., Cosic, K.: Comparative analysis of emotion estimation methods based on physiological measurements for real-time applications. International Journal of Human-Computer Studies 72, 717–727 (2014)
13. Felisberto, F., Fdez.-Riverola, F., Pereira, A.: A ubiquitous and low-cost solution for movement monitoring and accident detection based on sensor fusion. Sensors 14, 8961–8983 (2014)
14. Fernandez, R.: A Computational Model for the Automatic Recognition of Affect in Speech. Massachusetts Institute of Technology (2004)
15. Ambady, N., Rosenthal, R.: Thin slices of expressive behavior as predictors of interpersonal consequences: a meta-analysis. Psychological Bulletin 111, 256–274 (1992)
16. Zeng, Z., Pantic, M., Roisman, G.I., Huang, T.S.: A survey of affect recognition methods: audio, visual, and spontaneous expressions. IEEE Transactions on Pattern Analysis and Machine Intelligence 31, 39–58 (2009)
17. Lozano-Monasor, E., López, M.T., Fernández-Caballero, A., Vigo-Bustos, F.: Facial expression recognition from webcam based on active shape models and support vector machines. In: Pecchia, L., Chen, L.L., Nugent, C., Bravo, J. (eds.) IWAAL 2014. LNCS, vol. 8868, pp. 147–154. Springer, Heidelberg (2014)
18. Ekman, P., Friesen, W.V., O'Sullivan, M., Chan, A., Diacoyanni-Tarlatzis, I., Heider, K., Tzavaras, A.: Universals and cultural differences in the judgment of facial expressions of emotion. Journal of Personality and Social Psychology 53, 712–717 (1987)
19. Mauss, I.B., Levenson, R.W., McCarter, L., Wilhelm, F.H., Gross, J.J.: The tie that binds? Coherence among emotion experience, behavior, and physiology. Emotion 5, 175–190 (2005)
20. Van Galen, G.P., Müller, M.L., Meulenbroek, R.G., Van Gemmert, A.W.: Forearm EMG response activity during motor performance in individuals prone to increased stress reactivity. American Journal of Industrial Medicine 41, 406–419 (2002)
21. Coombes, S.A., Cauraugh, J.H., Janelle, C.M.: Emotional state and initiating cue alter central and peripheral motor processes. Emotion 7, 275–284 (2007)
22. Stemmler, G.: The autonomic differentiation of emotions revisited: convergent and discriminant validation. Psychophysiology 26, 617–632 (1989)

Part IV

Special Session on Web Mining and Recommender Systems (WebMiRes)

Feeding Software Agents with Web Information*

Patricia Jiménez[1], Hassan A. Sleiman[2], and Rafael Corchuelo[1]

[1] ETSI Informática, Avda. Reina Mercedes, s/n, Sevilla E-41012, Spain
{corchu,patriciajimenez}@us.es
[2] CEA, LIST Institute, Gif-sur-Yvette 91191 CEDEX, France
hassan.sleiman@cea.fr

Abstract. Many software agents require information that is available in web documents. Unfortunately, the existing proposals to learn extraction rules are tightly coupled with the learning component and do not result in resilient rules. We present a novel approach that leverages neural networks and has proven to be very resilient.

Keywords: Web information extraction, neural networks, resiliency.

1 Introduction

Some software agents need information that is provided by web sites, which is particularly problematic in cases in which there is not an API available. Our focus is on learning information extraction rules that help software agents in this task in the context of semi-structured web documents.

Surprisingly, few techniques in the literature have attempted to leverage general-purpose machine-learning techniques, but, instead, rely on ad-hoc learning procedures that cannot be easily decoupled from the rest of the proposal [3]; in other words, they cannot easily benefit from the many yearly advances in the field of machine learning. Furthermore, the existing techniques focus on learning very precise rules with high recall, which commonly makes them break when a web site changes. A few authors have paid attention to the problem of learning resilient rules [5, 6], but they are mostly handcrafted approaches. Other authors have focused on re-learning broken extraction rules as automatically as possible [10, 12, 13], but the problem is to detect that a rule is broken and that agents must stop until a new rule is available.

We present a novel approach to web information extraction, Dykers. It leverages neural networks to learn very resilient extraction rules; furthermore, it is decoupled from the core learner, which helps update it as new research results are published. The paper is organised as follows: Section 2 surveys the related work; Section 3 describes the details of our proposal; Section 4 reports on the results of our experiments; finally, Section 5 concludes our work.

* Supported by grants TIN2007-64119, P07-TIC-2602, P08-TIC-4100, TIN2008-04718-E, TIN2010-21744, TIN2010-09809-E, TIN2010-10811-E, TIN2010-09988-E, TIN2011-15497-E, and TIN2013-40848-R. Thanks to Dr. Antonia M. Reina for her help with a preliminary version of Dykers.

J. Bajo et al. (eds.), *Trends in Prac. Appl. of Agents, Multi-Agent Sys. and Sustainability*,
Advances in Intelligent Systems and Computing 372, DOI: 10.1007/978-3-319-19629-9_15

2 Related Work

Kushmerick et al. [11] pioneered this field with a proposal that learns token patterns that characterise the context of the information to extract; Hsu and Dung [8] devised a proposal that first learns an automaton that models the information to extract and then regular expressions to model transitions; Hogue and Karger [7] presented a proposal that is based on tree similarity; Álvarez et al. [1] devised a proposal that relies on clustering, tree matching, string matching, and string alignment; Crescenzi and Merialdo [4] presented a proposal to infer a regular expression that models the differences amongst a number of documents, which are typically the information of interest; Kayed and Chang [9] devised a technique to learn rules that are context-free grammars; and Sleiman and Corchuelo [14, 15] presented two proposals that are based on multi-string alignment.

The previous proposals rely on techniques that were specifically tailored to learn information extraction rules. That is, the core learning component is not a standard machine-learning technique and it is not clearly differentiated from the rest of the proposal, which makes it difficult to update it with the new advances in machine learning. Contrarily, Dykers leverages neural networks, but keeps them decoupled from the rest of the proposal.

An additional problem is that the rules learnt are not typically intended to be resilient. This motivated Embley [5] or Gregg and Walczak [6] to work on proposals that require the user to devise an ontology that acts as a high-level extraction rule, which is expected to be more resilient than an automatically-learnt rule. Other authors have focused on procedures to re-learn broken rules, which seems easier [10, 12, 13]. They learn a model from correct information and then check whether a piece of information extracted by a rule deviates significantly from that model, in which case an alarm is raised; then a human expert must assess the results and, if necessary, run a re-learning process that attempts to re-use as much information from the original training set as possible.

Dykers can learn very resilient extraction rules. The user just provides a training set with examples of the information required to feed an agent and it uses a novel generalisation and mutation procedure that combined with a novel search procedure helps learn very resilient rules very efficiently.

3 Our Proposal

In this section, we describe the main procedure in Dykers, then the procedure to learn a rule, and, finally, the scoring function that guides the search process.

3.1 Main Procedure

Figure 1 shows the main procedure of Dykers. It works on a collection of datasets sc. For each dataset ds in sc, it repeats the following procedure β times: first, it selects $\lceil \gamma |ds| \rceil$ documents from dataset ds and then δ documents from each of the remaining datasets, where β, γ, and δ are user-defined parameters. This

```
1: Dykers(sc)
2:    m = ∅
3:    for each dataset ds in sc do
4:       rs = ⟨⟩
5:       repeat β times
6:          ds1 = select ⌈γ |ds|⌉ documents from ds
7:          ds2 = select δ documents from each dataset in sc, except for ds
8:          ts = compute training set for ds1 ∪ ds2
9:          vs = compute validation set for ds \ ds1
10:         ts = generalise and mutate ts
11:         (u, v, n, s) = learnRule(ts, vs)
12:         insert (u, v, n, s) in rs with priority s
13:      end
14:      m = m ∪ {(ds, rs)}
15:   end
16: return m
```

Fig. 1. Main procedure of Dykers

Table 1. Feature generalisations

Feature	Generalisation
Coordinates	The coordinate system is split into squares that are 10% high and 10% wide the size of the page when it is rendered. Coordinates are transformed from nominal values onto the corresponding disecretised values.
Tag names	Tag names are generalised as format tags (<b>, <i>, <emph>, <strong>, and the like), meta tags (<meta>, <script>, <style>, and the like), and structural tags (<p>, <table>, <div>, and the like).
Font sizes	Font sizes are generalised as unimportant (which includes sizes that are classified by the browser as xxsmall, xsmall, or small), normal (which includes sizes that are classified as such by the browser, with a ±10% tolerance), and highlighted (which includes large, xlarge, and xxlarge sizes).
Font weights	Font weights are generalised as normal or emphasised (bold, bolder, emphasis, underlined, and the like).
Colours	Colours are generalised as normal (the most common one) and emphasis (the other colours).
Widths	Widths are classified as normal (the most common), unimportant (those that are 25% thinner than the normal widths), and emphasised (those that are 25% thicker than the normal widths).

helps create a combined training set in which most documents are expected to have been downloaded from the same site, but there is some noise from other sites that has proved to help learn very resilient rules.

The training and the validation sets are then computed, which basically consist in generating a vector-based representation of the selected documents. Each vector corresponds to a DOM node; the first component is its unique identifier (which is obviously ignored for learning purposes), then come the values of its features, and then its class. The set of features includes every standard HTML and rendering attribute, plus a number of user defined features like the first bigram, the ratio of letters, the number of tokens, and the like.

Then, the training set is generalised, which consists in changing the values of some features so that they become more general. For instance, a DOM node may have font size '26pt' and another may have font size '24pt'; using these exact values leads to a rule that is very specific and less resilient than a rule that specifies that the font size is large. Similarly, it does not actually matter if a piece of text is underlined or bold, what matters is that it is emphasised. Table 1 summarises the generalisations that Dykers performs.

```
1: learnRule(ts, vs)
2:     n = learn network from ts
3:     s = score(n, vs)
4:     r = r' = (ts, vs, n, s)
5:     repeat
6:         c = computeCandidates(r')
7:         if c ≠ ⟨⟩ then
8:             r = r'
9:             r' = refineRule(r', c)
10:        end
11:    until c = ⟨⟩ or r = r'
12: return r'
```

Fig. 2. Learning a rule

Next, the generalised training set is mutated, which consists in changing the values of some features randomly or making changes to the structure of a document. Dykers performs the following structural changes: remove parent nodes, remove children nodes, remove left or right sibling, exchange a node and its parent, exchange a node and its children, exchange a node and its siblings.

Then a rule is learnt from the resulting generalised and mutated training set; rules are of the form (u, v, n, s), where u denotes the training set from which it was learnt, v is a validation set, n is a neural network that encodes the information in u, and s is the score of n when applied to validation set v.

Recall that Dykers repeats the previous procedure β times. In each iteration, it learns a rule from a different dataset; these rules are stored in a queue rs using their scores as the priority. When a new node is to be classified as providing information to be extracted or not, all of the rules are applied and the majority class is returned. The main procedure then learns β extraction rules for each of the datasets that is passed as a parameter; it returns a map in which each dataset is associated with its corresponding collection of extraction rules.

3.2 Learning a Rule

Figure 2 shows the procedure to learn a rule from a training set ts using validation set vs. It first computes an initial rule (ts, vs, n, s), where n is a neural network learnt from ts and s denotes its score on validation set vs. We do not commit to a particular kind of neural network; Dykers can leverage any proposal in the literature [2], but our experiments prove that RBFN networks work very well in this context. Then, it iteratively computes candidate expansions of the current rule and uses them to refine it until no expansion is possible.

The left part of Figure 3 shows the procedure to compute the candidate expansions of a rule r. It first decomposes r and then computes the expansions of its training set. Computing such expansions is simple: it creates a new training set in which the vectors, each of which corresponds to a DOM node, have additional components that correspond to the features of a neighbour. For instance, Figure 4 shows an excerpt of an initial training set on the left, which provides information about the nodes in the input documents; on the right, there is an

1: *computeCandidates*(r)
2: (u, v, n, s) = components of r
3: e = compute expansions of u
4: $c = \langle\rangle$
5: for each u' in e do
6: n' = learn network from u'
7: v' = expand v according to u'
8: $s' = score(n', v')$
9: insert (u', v', n', s') in c with priority s'
10: end
11: return c

1: *refineRule*(r, c)
2: (u, v, n, s) = components of r
3: r' = remove the head of c
4: (u', v', n', s') = components of r'
5: while $s' > s$ do
6: $r = r'$
7: (u, v, n, s) = components of r
8: if $c \neq \langle\rangle$ then
9: r' = remove the head of c
10: (u', v', n', s') = components of r'
11: u' = combine u and u'
12: v' = combine v and v'
13: n' = learn a network from u'
14: $s' = score(n', v')$
15: $r' = (u', v', n', s')$
16: end
17: end
18: return r

Fig. 3. Procedures to compute candidate expansions and best candidates

node	font Family node	font Size node	tag node	class node
n1	Times	14px	a	null
n2	Arial	16px	div	book
n3	Null	null	img	null
n4	Times	14px	a	title
n5	Helvetica	12px	strong	null
n6	Helvetica	12px	null	null
n7	Verdana	12px	a	author
n8	Helvetica	13px	a	author

node	font Family node	font Size node	tag node	node1 = parent (node)	font Family node1	font Size node1	tag node1	class node
n1	Times	14px	body	null	null	null	null	null
n2	Arial	16px	div	n1	Times	14px	a	book
n3	Null	null	img	n2	Arial	16px	div	null
n4	Times	13px	a	n2	Arial	16px	div	title
n5	Helvetica	12px	span	n2	Arial	16px	div	null
n6	Verdana	12px	null	n5	Courier	12px	span	null
n7	Verdana	12px	a	n2	Arial	16px	div	author
n8	Verdana	12px	a	n2	Arial	16px	div	author

Fig. 4. Excerpt of an expanded training set

expanded training set in which each vector has been extended with the features of its parent. Dykers takes into account the following neighbouring relationships: parent, left sibling, right sibling, and i-th child, for $i = 1..m$, where m represents the maximum number of children of a node in the collection of input documents. Then, the procedure iterates through the set of expansions; for each expansion u' it learns a neural network n', computes the appropriate validation set v', computes its score s', and then stores the result in a priority queue c that is returned once every expansion has been analysed.

The right part of Figure 3 shows the procedure to refine a rule. It works on the rule r that has been learnt so far and the set of candidates c that was computed previously. It first decomposes rule r into its components and then removes the first candidate r' and decomposes it. Then, it iterates as long as the score of r' is higher than the score of r, i.e., it provides some gain; recall that the collection of candidates is sorted in descending order of score, so the first one is the one that corresponds to exploring the neighbour that provides the maximum score. In each iteration, it removes the first remaining candidate from c and combines its

Table 2. Dataset used in our experimentation

Books	Movies	Cars	Doctors	Jobs	Real estate
Abe Books	Disney Movies	Auto Trader	Web MD	Insight into Diversity	Haart
Awesome Books	Albania Movies	Car Max	Ame. Medical Assoc.	4 Jobs	Homes
Better World Books	All Movies	Car Zone	Dentists	6 Figure Jobs	Remax
Many Books	CITWF	Classic Cars for Sale	Dr. Score	Career Builder	Trulia
Waterstones	Soul Films	Internet Autoguide	Steady Health	Job of Mine	Yahoo

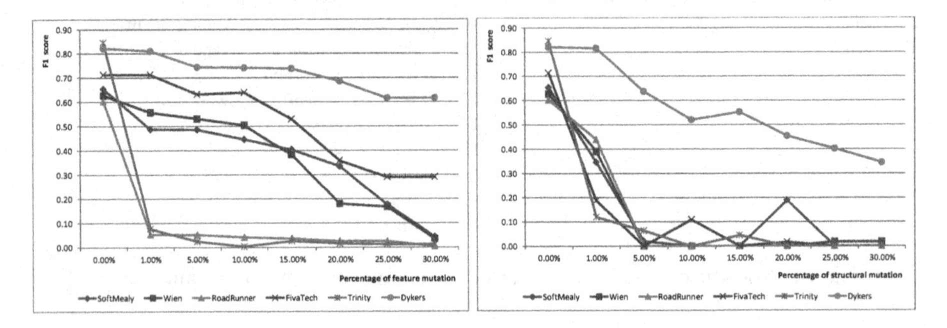

Fig. 5. Experimental results

training and validation sets with the training and validation sets of the current rule, which helps explore as many neighbours as possible in a single iteration.

3.3 Scoring Rules

Dykers relies on the following scoring function to assess how good a rule is:

$$\frac{2\,h\,m}{h+m},$$

where h denotes the fraction of correct information extracted (hits) and m denotes one minus the fraction of incorrect information extracted (misses). Both figures are averaged using a harmonic mean; that is, it rewards a rule when its number of hits is high and the number of misses is low, and penalises it otherwise.

To compute h and m, we have to use a rule's neural network to predict the class of every vector in a validation set and then count the number of true positives (tp), true negatives (tn), false positives (fp), and false negatives (fn). Fractions h and m are then defined as follows:

$$h = (tp + tn)/(tp + tn + fp + fn)$$
$$m = 1 - (fp + fn)/(tp + tn + fp + fn).$$

4 Experimental Results

We performed our experiments on a four-threaded Intel Core i7 computer that ran at 2.93 GHz, had 4 GiB of RAM, Windows 7 Pro 64-bit, Oracle's Java

Development Kit 1.7.9_02, and Weka 3.6.8. The experiments were performed on a collection of documents that were downloaded from the sites in Table 2; from each site, we downloaded 30 documents that were selected randomly.

We searched the web and contacted many authors in order to have access to the implementation of as many proposals as possible. We managed to find an implementation for SoftMealy [8] and Wien [11], which are classical proposals, and RoadRunner [4], FiVaTech [9], and Trinity [15], which are recent proposals.

Regarding Dykers, we experimented with several combinations of parameters and kinds of neural networks. We found out that the following values for the parameters work quite well: $\beta = 10$, that is, 10 rules are learnt for each site; $\gamma = 0.30$, that is, 30% of the documents available in each site are used for training purposes and the remaining 70% documents for validation purposes; and $\delta = 1$, that is, one document is selected from sites other than the one for which Dykers is learning a rule. Regarding the learning technique, we found out that RBFN networks [2] are the best performing in this context.

We ran Dykers and the other proposals on the previous datasets and computed precision and recall, which were aggregated using the F_1 score. We first ran the proposals on the original datasets and then introduced random mutations to 1%, 5%, 10%, 15%, 20%, 25%, and 30% of the nodes. Figure 5 shows the results, which make it clear that Dykers learns rules that are very precise and have high recall when no changes are involved, but keep working better than the other proposals when the documents undergo changes.

Regarding feature mutation, realise that neither RoadRunner nor Trinity perform well since a mutation as light as 1% has a dramatic impact on their performance. The reason is that both techniques are intended to learn a regular expression that provides a template for a whole document; such expressions rely on the tags, their HTML attributes, and the text; thus a small change results in a totally broken extraction rule. Contrarily, SoftMealy and Wien are a little more resilient because they only characterise the tokens that surround the information to be extracted; as far as a mutation does not change those tokens, they can keep extracting information. Note, however, that their performance is poor in general because the Web and the way that HTML is used has varied significantly since these proposals were published. FiVaTech, on the contrary, seems to be the most resilient proposal after Dykers; it also works on DOM trees, whose nodes are clustered according to a similarity function and then generalised into tree patterns; this results in rules that can deal with feature changes more easily than other proposals, but worse than Dykers.

Regarding structure mutation, all of the techniques are clearly negatively affected, but Dykers still produces the most resilient extraction rules. The results regarding RoadRunner or Trinity are not surprising at all since small changes to the HTML totally break the regular expressions that they learn. Regarding SoftMealy and Wien, they are also very clearly affected by structure mutation because a change to the structure of the DOM tree results in a reorganisation of the tokens that surround the information to be extracted, which quickly breaks its extraction rules. That is the reason why FiVaTech is also very affected by structural changes since the tree patterns that it learns are very dependent on the nested structure of the DOM nodes.

5 Conclusions

We have introduced Dykers, which is a new proposal to learn web information extraction rules. It can easily benefit from the innovations developed by other authors since the core learning technique can be viewed as a plug-in component. Our analysis of the related work proves that Dykers is innovative from a conceptual point of view, since it clearly deviates from current approaches in the literature; furthermore, our experimental analysis proves that it produces very resilient rules when RBFN neural networks are used. This makes it an excellent choice to feed software agents with web information.

References

[1] Álvarez, M., Pan, A., Raposo, J., Bellas, F., Cacheda, F.: Extracting lists of data records from semi-structured web pages. Data Knowl. Eng. 64(2), 491–509 (2008)

[2] Bianchini, M., Maggini, M., Jain, L.C.: Handbook on Neural Information Processing, vol. 49. Springer (2013)

[3] Chang, C.H., Kayed, M., Girgis, M.R., Shaalan, K.F.: A survey of web information extraction systems. IEEE Trans. Knowl. Data Eng. 18(10), 1411–1428 (2006)

[4] Crescenzi, V., Merialdo, P.: Wrapper inference for ambiguous web pages. Applied Artificial Intelligence 22(1&2), 21–52 (2008)

[5] Embley, D.W.: Towards semantic understanding – an approach based on information extraction ontologies. In: ADC, p. 3 (2004)

[6] Gregg, D.G., Walczak, S.: Exploiting the Information Web. IEEE Transactions on Systems, Man, and Cybernetics, Part C 37(1), 109–125 (2007)

[7] Hogue, A.W., Karger, D.R.: Thresher: Automating the unwrapping of semantic content from the World Wide Web. In: WWW, pp. 86–95 (2005)

[8] Hsu, C.N., Dung, M.T.: Generating finite-state transducers for semi-structured data extraction from the web. Inf. Syst. 23(8), 521–538 (1998)

[9] Kayed, M., Chang, C.H.: FiVaTech: Page-level web data extraction from template pages. IEEE Trans. Knowl. Data Eng. 22(2), 249–263 (2010)

[10] Kushmerick, N.: Regression testing for wrapper maintenance. In: AAAI/IAAI, pp. 74–79 (1999)

[11] Kushmerick, N., Weld, D.S., Doorenbos, R.B.: Wrapper induction for information extraction. In: IJCAI (1), pp. 729–737 (199 7)

[12] Lerman, K., Knoblock, C.A.: Wrapper maintenance. In: Encyclopedia of Database Systems, pp. 3565–3569. Springer (2009)

[13] McCann, R., AlShebli, B.K., Le, Q., Nguyen, H., Vu, L., Doan, A.: Mapping maintenance for data integration systems. In: VLDB, pp. 1018–1030 (2005)

[14] Sleiman, H.A., Corchuelo, R.: TEX: An efficient and effective unsupervised web information extractor. Knowl.-Based Syst. 39, 109–123 (2013)

[15] Sleiman, H.A., Corchuelo, R.: Trinity: On using trinary trees for unsupervised web data extraction. IEEE Trans. Knowl. Data Eng. 26(6), 1544–1556 (2014)

A Knowledge-Based Recommender Agent to Choosing a Competition System

Vivian F. López, Rubén E. Salamanca, María N. Moreno, Ana B. Gil, and Juan M. Corchado

Departamento Informática y Automática, University of Salamanca, Plaza de la Merced S/N, 37008. Salamanca
{vivian,u76383,mmg,abg,corchado}@usal.es

Abstract. In this paper, we present a knowledge-based recommender agent, which is part of a multiagent system (MAS) that controls all the basic tasks involved in organizing a sporting tournament. Using a recommendation algorithm, this agent finds one competition system that best suits the competitions characteristics in organizing a tournament. We show how knowledge-based recommender systems can be used to guide sport managers in the choice of competition system. This provides an interesting decision alternative, based on in the needs of the users and tournament data.

1 Introduction

The most important part in the development of any sporting competition is the organization. Many aspects must be considered, such as the selection of infrastructure, the selection of a competition system, advertising, regulations, distribution of information between participants, and so on, which can often become a tedious task, or can even cause problems during the course of the event. Given the wide range of existing sports, each with its own set of rules, there are different ways to compete. Some sports are played individually (athletics), others in pairs (tennis) and in teams (football or basketball). This leads to the existence of various systems of competition as seen below.

In this paper, we will show how to apply a knowledge-based recommender system to guide sport managers in their choice of a competition system. We present a recommender agent, which is part of a MAS. We believe that the use of this agent will improve the management of a sporting event regarding all the basic tasks involved in organizing a tournament: the selection of the sports venue, registration, reporting mechanisms, tracking, storing and diffusing information for each event in real time.

This paper is structured as follows. In Section 2 we present the recommender systems, section 3 provides a short overview of the knowledge-based recommender agent, in which a recommendation algorithm is intended to guide sport managers in the choices made for a competition system. This algorithm finds the system that best suits the competitions characteristics. Section 4 explains the conclusions of the paper and states future work.

J. Bajo et al. (eds.), *Trends in Prac. Appl. of Agents, Multi-Agent Sys. and Sustainability*,
Advances in Intelligent Systems and Computing 372, DOI: 10.1007/978-3-319-19629-9_16

2 Recommender System

A recommender system is a program that analyses behaviors and characteristics of users and attempts to recommend actions or items that could be useful to those users [6], [18]. This type of system began to appear in the market in 1996 [11]. Applying the idea of composing profiles from the behaviors, preferences and characteristics of users, recommender systems can be used, based on different areas of knowledge, to learn about user interests and to make recommendations accordingly [4].

The main types of recommendation systems are content-based, collaborative filtering, or knowledge-based or they can be hybrid recommender systems, which combine two or more techniques to improve the quality of recommendations [1][12].

In the content based approach, items are recommended according to their similarities to items sought by users in the past. Items are recommended by comparing their content and user profiles [10].

Collaborative filtering techniques predict product preferences for a user based on the opinions of other users. The opinions can be obtained explicitly from the users as a rating score or by using some implicit measures from purchase records, such as timing logs [17]. Currently there are two approaches for collaborative filtering, memory-based (user based) and model-based (item-based) algorithms. Memory-based algorithms, also known as nearest-neighbour methods, were the earliest used of the two [14]. They treat all user items by means of statistical techniques in order to find users with similar preferences (neighbours). The advantage of these algorithms is the quick incorporation of the most recent information; however, the inconvenience is that the search for neighbours in large databases is slow [16].

Model-based collaborative filtering algorithms use data mining techniques in order to develop a model of user ratings, which is used to predict user preferences. Collaborative filtering, especially the memory-based approach, has some limitations in the e-commerce environment. Sparsity and scalability are serious weaknesses which would lead to poor recommendations [5]. Sparsity occurs when the number of ratings needed for prediction is greater than the number of the ratings obtained, and may take place because collaborative filtering usually requires user explicit expression of personal preferences for products. The second limitation is related to performance problems in the search for neighbours in memory-based algorithms. These drawbacks may be minimized by means of data mining methods. However, there are other shortcomings that may occur even with these methods. The first-rater (or early-rater) problem arises when it is impossible to offer recommendations about an item that was just incorporated into the system and, therefore, has few, if any, evaluations from users. Analogously, such drawbacks also occur with a new user joining the system, since there is no information about them. It would be impossible to determine the user's behavior in order to provide recommendations. Currently, this variant of the first-rater problem is also referred to as the cold-start problem. Semantic web mining can be used to address the last problems. Taxonomic abstraction provided by an ontology allow for inducing pat- terns at a more abstract level; that is, regularities can be found between categories of products instead of between specific products. These patterns can be used in recommender systems for recommending new products that still have not been rated by the users [9]. This is a way of dealing with the first-rater problem. The cold-start problem can be solved in a similar way.

In summary, we can say that the traditional systems of recommendations, such as collaborative filtering or based on content, rely on user profiles and certain features of the items to make recommendations. However, these systems do not explore in depth knowledge about the domain of items. Therefore, the systems traditional recommendations are perfectly valid for processes recommendation of unique products, but they are not are useful in complex domains. At present there is an interesting approach, which has emerged as a trend in the next generation of recommender systems: to contemplate user profiles by recognizing what users need and not simply what they want or prefer [13]. In this approach, as pointed out in [6], recommender systems are more useful to users because they can help them to solve problems, make decisions, streamline tasks and even transform their profiles. Thus, we believe that knowledge-based recommender systems would be more suitable to guide sport managers in the choice of competition system.

3 Knowledge-Based Recommender Systems

A third type of recommender system is one that uses knowledge about users and products to pursue a knowledge-based approach to generating a recommendation, by using reasoning to determine which products meet the users requirements [2]. They model the user profile and apply inference algorithms, in order to identify the correlation between user preferences and existing products, services or content [3].

These systems aim to deepen the knowledge about users and items in order to make recommendations that properly fit with the requirements of the user. Based on this type of information, the system performs a reasoning process that fits more items associated with the needs of the user. Thus, the key to the knowledge-based recommender systems lies in a more thorough understanding of how the user defines a need. In [2][3][7] we can see several examples of knowledge-based recommender systems.

The operating mode of knowledge-based systems may lead to the elimination of certain problems generally associated with recommendation systems:

1. *Cold-start* problems may be eliminated, as their recommendations do not depend on user ratings.
2. There is no need to store information about a particular user because similarities between user preferences are independent of each other.

Furthermore, such recommendation systems have a number of drawbacks:

1. A knowledge engineering is required.
2. The recommendations are static type, i.e., for the same cases, the same recommendations are made. This does not happen with recommender systems based on collaborative filtering because they adapts to the recommendations as more information about the user is learned.

3.1 Knowledge-Based Recommender Agent

Given the number of competition systems that currently exist and the advantages and disadvantages of each of them, selecting one competition system is one of the most difficult decisions for managers to make. However, the traditional systems of recommendation are not the most appropriate to solve this problem, since the preferences of other users or past performance provides no valid or sufficient information to recommend competition systems. Therefore, we think recommendation systems based on knowledge are more appropriate because they use information provided by the user about specific needs and tournament data. With this information the system makes the best recommendations for the managers.

For this reason, we decided to implement an agent capable of guiding a manager in choosing a competition system that best suits their tournament needs and/or preferences. A model of a single system for efficient competitions management has many limitations. The system would need an enormous amount of knowledge to be able to effectively deal with user information requests that cover a variety tasks. For reasons, in addition to the characteristics of the Internet environment, we employ a distributed collaborative collection of agents to draw up an optimum competition system.

The algorithms used in the knowledge-based systems are founded in case-based reasoning (CBR) [8]. They solve new problems with a high capacity for learning and adaptation. There are three different types of knowledge [15]:

1. Catalogue knowledge: This is the knowledge that we have of the recommending items and its features.
2. Functional knowledge: This is the understanding of how the recommending items can match the needs of users.
3. User's knowledge: We need to gather information about the needs of the user in order to find items that meet their needs.

In order to develop the knowledge-based recommender agent, we have built an algorithm based on these principles, which are detailed below. The recommendation algorithm (Algorithm1) is based on assigning a score to each competition system and recommending the system that has gained the highest score at the end of the process.

3.2 Recommendation Algorithm Based on Scoring

As the recommendation system based on knowledge is implemented, the algorithm starts from a set of competition systems S which comprises the products to recommend, a set of base rules R, and a set of user preferences P. Both sets R and P form the set of evaluation rules E. Each rule is given a weight k_n so that we establish an order of importance for each rule in the set of evaluation. As a result of executing the algorithm, the competition system with the highest score is recommended.

Algorithm 1. Recommendation algorithm based on scoring

1: **Input**: S set of competition systems (products to recommending), $S = \{s_1, s_2, ...s_n\}$, P is the set of user preferences $P = \{p_1, p_2, ...p_n\}$, R is the set base rules $R = \{r_1, r_2, ...r_n\}$, and E is the set of evaluation rules, $E = (R \cup P) = \{e_1, e_2, ...e_n\}$.
2: **Output**: Competition system recommended
3: **for** all e_n of E **do**
4: $e_n = k_n$
5: compute resulting score v of s_n defined as

$$v = \begin{cases} 1 & \text{if the system is optimal for the rule.} \\ 0 & \text{if the system is not optimal for the rule.} \\ \text{-a/k} & \text{if the system is bad for the rule.} \end{cases}$$

where a is cumulative score for competition systems s_n, defined as $a = a + v * k$ (score calculating function).
6: **for** all s_n of S **do**
7: initialize $a_n = 0$
8: **end for**
9: sort set of evaluation rules E, according to their weights of manner descending.
10: **for** all e_n of E **do**
11: initialize $a_n = 0$
12: **for** all s_n of S **do**
13: the rule e_n is applied to obtaining the value of e.
14: new resulting score is assigned based on step 5.
15: return competition system with the highest score (compute resulting score v).
16: **end for**
17: **end for**
18: **end for**

The organizer user must finally decide which competition system to use and, on that basis, determine the calendar of matches. The agent provides the option Calendar / Matches, from which the organizer may determine the system that will govern their competition. By clicking on the competition system button, a wizard appears, allowing the user organizer to choose these options:

1. Recommended which competition system is best adapted to the characteristics of the tournament or, conversely,
2. Manually select the competition system to use.

If an organizer user chooses to accept the system recommendation, an interface will appear in which the user can input specific preferences.

Once the agent has assessed all the rules, the element whose score resulting was higher is recommended, in this case the Mixed competition system. The recommender agent represents a helpful tool to guide sport managers in the choice of competition system. From the MAS, recommender agent collaborates with the tournaments agent to allow any organizer user to register a tournament. It will facilitate planning, in addition to publicizing the tournament and the registration of the participating teams.

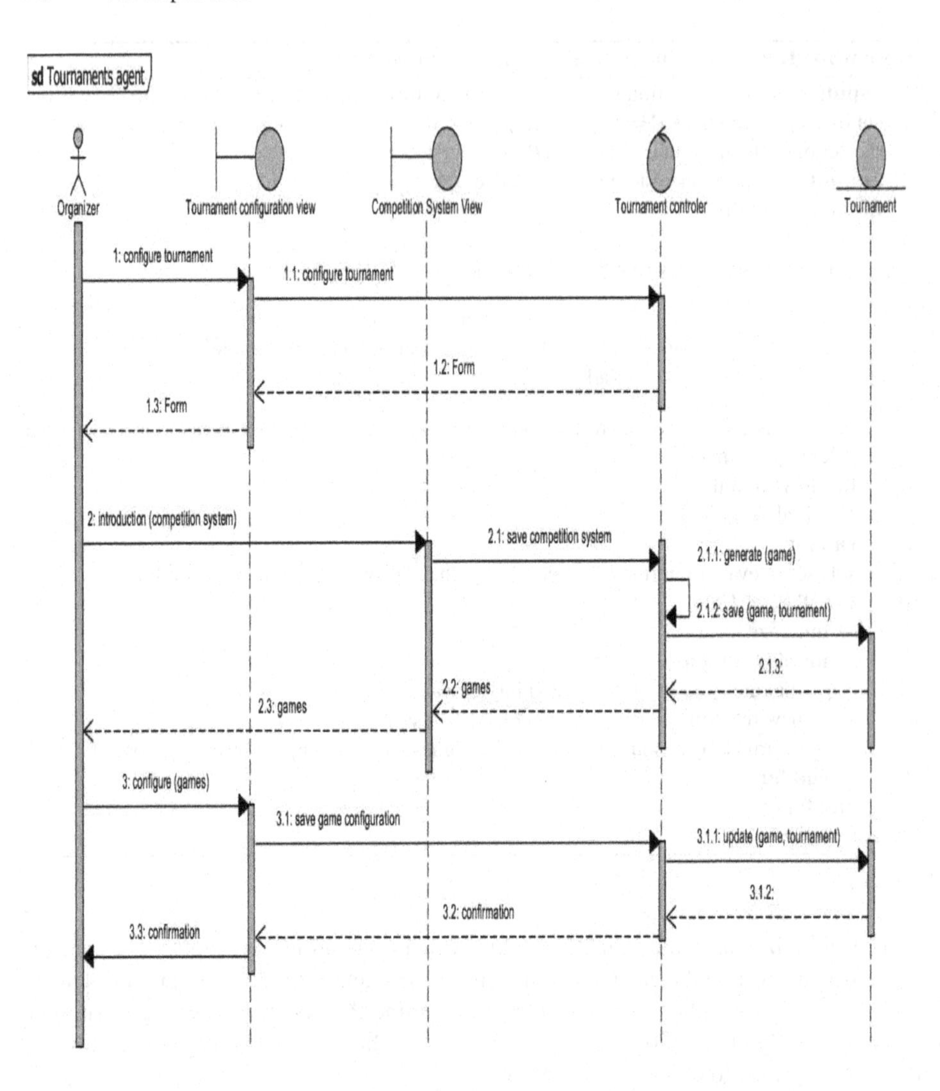

Fig. 1. The sequence diagram tournaments Agent

In Figure 1 we can see the sequence diagram applied to define the external view point to the tournaments agent, which represents the model interactions in various phases of software development.

4 Conclusions

The use of MAS for managing a sporting events can improve the performance in the organization of a competition. The managers of sporting events can alleviate or delegate responsibilities and tasks in an automated system and improve information disclosure between the various participants in real time. For this reason, we decided implement an agent capable of guiding an organizer through the process of choosing a competition system that suits their tournament needs and/or preferences. This agent uses a recommendation algorithm based on scoring, which has made it possible to build a knowledge-based recommender system in the sporting context, while the typical use of such a system is part of electronic commerce. The result obtained has been better than expected, resulting in a recommender system that can assist tournament organizers in selecting the most suitable competition system. The model offers a methodology for the efficient organization of sporting events, becoming a more appropriate tool for management level decision making. As future work we will improve the functionality of the agent using other data mining techniques to improve the algorithms used in recommender systems. We would also like to include features that allow organizers to disseminate information about tournaments through social networks, and to know the popularity of their tournaments or the success of the dissemination efforts. This could be accomplished by applying data mining and opinion mining techniques.

Acknowledgments. This paper has been partially supported by the MINISTERIO DE ECONOMÍA with the Research Project TIN2012-36586-C03-03, "*Sociedades Humano-Agente: Inmersión, Adaptación y Simulación*".

References

1. Adomavicius, G., Tuzhilin, A.: Toward the next generation of recommender systems: A survey of the state-of-the-art and possible extensions. IEEE Transactions on Knowledge and Data Engineering 17, 734–749 (2005)
2. Burke, R.: Knowledge-based recommender systems. In: Encyclopedia of Library and Information Systems (2000)
3. Carrer-Neto, W., Hernández-Alcaraz, M.L., Valencia-García, R., García-Sánchez, F.: Social knowledge-based recommender system. Application to the movies domain. Expert Systems with Applications 39, 10990–11000 (2012)
4. Chen, R., Huang, Y., Bau, C., Chen, S.: A recommendation system based on domain ontology and SWRL for anti-diabetic drugs selection. Expert Systems with Applications 39, 3995–4006 (2012)
5. Cho, H.C., Kim, J.K., Kim, S.H.: A personalized recommender system based on web usage mining and decision tree induction. Expert Systems with Applications 23, 329–342 (2002)
6. Gonçalves, M., Marques, P., Ciarelli, O.E.: Recommendation of programming activities by multi-label classification for a formative assessment of students. Expert Systems with Applications 40, 6641–6651 (2013)
7. Hayes, C., Cunningham, P.: Smart radio - community based music radio. Knowledge-Based Systems 14(3-4), 197–201 (2001)
8. Hammond, K.: Case-base planning: Viewing Planning as a Memory Task. Academic Press, Boston (1989)
9. Huang, Y., Bian, L.: A Bayesian network and analytic hierarchy process based personalized recommendations for tourist attraction over the Internet. Expert Systems with Applications 36, 933–943 (2009)
10. Lee, C.H.,, Kim, Y.H., Rhee, P.K.: Web personalization expert with combining collaborative filtering and association rule Mining Technique. Expert Systems with Applications 21, 131–137 (2001)

11. Manber, U., Patel, A., Robison, J.: Experience with personalization of Yahoo! Communications of the ACM 43(8), 35–39 (2000)
12. Manouselis, N., Vuorikari, R., Van Assche, F.: Collaborative recommendation of e-learning resources: An experimental investigation. Journal of Computer Assisted Learning 26, 227–242 (2010)
13. Park, D.H., Kim, H.K., Choi, I.Y., Kim, J.K.: A literature review and classification of recommender systems research. Expert Systems with Applications 39, 10059–10072 (2012)
14. Resnick, P., Lacovou, N., Suchack, M., Bergstrom, P., Riedl, J.: Grouplens: An open architecture for collaborative filtering of netnews. In: Proc. of ACM Conference on Computer Supported Cooperative Work, pp. 175–186 (1994)
15. Ruan, D., Hardeman, F., Van Der Meer, K.: Intelligent Decision and Policy Making Support Systems. SCI, vol. 177. Springer, Heidelberg (1984)
16. Schapire, R.E., Singer, Y.: Boostexter, A boosting-based system for text categorization. Machine Learning 39(2/3), 135–168 (2000)
17. Schafer, J.B., Konstant, J.A., Riedl, J.: E-Commerce Recommendation Applications. Data Mining and Knowledge Discovery 5, 115–153 (2001)
18. Zaiane, O.: Building a recommender agent for e-learning systems. In: Proceedings of the International Conference on Computers in Education, vol. 1, pp. 55–59 (2002)

Combining Users and Items Rankings for Group Decision Support

Silvia Rossi[1], Antonio Caso[2], and Francesco Barile[2]

[1] Dipartimento di Ingegneria Elettrica e Tecnologie dell'Informazione, Universita' degli Studi di Napoli "Federico II", Napoli, Italy
silvia.rossi@unina.it

[2] Dipartimento di Fisica, Universita' degli Studi di Napoli "Federico II", Napoli, Italy
{antonio.caso,francesco.barile}@unina.it

Abstract. Traveling and city sightseeing are, in most cases, activities that involve small groups of users. Hence, a content personalization process, in a travel domain, requires taking into account simultaneously the preferences of different users. Moreover, a group recommendation system should also capture the possible intra-group relationships, which are fundamental features in a group decision process. In this paper, we model this problem as a multi-agent aggregation of preferences by using weighted social choice functions. In this context, weights can be extracted by analyzing the interactions of the group's members on Online Social Networks.

Keywords: Group Decision Making, Social Choice, Group Recommendation, Small Groups, Social Networks.

1 Introduction

A decision support system for planning a city tour has to take into account that, potentially, different group's members (not a single user) jointly select the activities to perform. Hence, it is necessary to suggest activities, or, more generally speaking, certain Points of Interest (POI) that maximize the group satisfaction, considering that the members' preferences can be different.

Providing recommendation to groups is a challenging problem because of the diversity and the dynamics of intra-group relationships [1]. The design and implementation of group recommendation systems, and, more generally, of decision support systems, should take into account the type of control in the group decision-making process, and consider the social relationships among the group members, intra-group roles and mutual influences. For example, the decision of a group member whether or not to accept a given recommendation may depend not only on his/her own evaluation of the content of the recommendation, but also on his/her beliefs about the evaluations of the other group members [2]. Users may be willing to accept last preferred activities in order to improve the group happiness.

Online Social Networks (OSNs) are widely recognized as effective ways to interact, communicate and collaborate with friends, but also to drive people's opinions. Moreover, OSNs interaction analysis can provide a viable way to obtain, without intruding

J. Bajo et al. (eds.), *Trends in Prac. Appl. of Agents, Multi-Agent Sys. and Sustainability*,
Advances in Intelligent Systems and Computing 372, DOI: 10.1007/978-3-319-19629-9_17

the users with questionnaires, information about the social relationships and pattern of activities among the group's members.

In this work, we show how social choice functions, typically used to merge agents' preferences, can be adopted to provide group recommendations. Moreover, an automatic analysis of group relationships in OSN can be used to weigh or to provide a rank of the group members, in order to include social roles and mutual influences in such functions. To validate our approach, we conducted two user studies with groups involved in a city trip planning. In the first, each user was asked to select his/her preferred POI, from a list, and then the group to provide a final decision. In the second, the users were asked to provide the rankings for the same POI and than to evaluate two solutions for the groups that are recommended by the system.

2 Related Works

The problem of aggregating individual preferences has been widely studied in Mathematics, Economics and Multi-agent systems (MAS), with the definition of *Social Choice* functions. These strategies, according to [3], can be classified as majority-based (mainly implemented as voting mechanisms to determine the most popular choices among alternatives), consensus-based (that try to average among all the possible choices and preferences), and role-based (that explicitly take into account possible roles and hierarchical relationships among members). PolyLens [4] has been one of the first approaches to include social characteristics (such as the nature of a group, the rights of group members, and social value functions for groups) within the group recommendation process. Other works try to address the problem with other classical MAS approaches, like *Negotiation* and *Game Theory*. For example in [5] the authors use different negotiation approaches, with agents that act on behalf of group's members, and the recommendations are obtained as result of the negotiation, while in [6] the authors try to apply game theory techniques, modeling the problem as a game and determining the recommendation by finding the nash equilibrium of the game.

However, many of these approaches do not consider the social relationships among group members at all [1]. Following this idea, the Gartrell's work [1] starts to evaluate the group members' weights, in terms of their importance or influence in a group, for movies recommendations. The scope of this work is similar to ours and it was tested on real groups. However, in [1] the defined group functions rely on the concept of "expertise" and "group dissimilarity", while the selection of the function is obtained by a "social value" derived from questionnaires. On the contrary, in our work, we argue that relevant parameters can be directly extracted, in a first approximation, by a "non-semantic" analysis of the interactions and by using an extension of widely used social choice functions. Finally, consensus decision-making has been explored, inspired by social networks, in [7], where individual preferences are aggregated in a weighted social choice function that, as in our case, takes into account local relationships with neighborhoods in a network. However, in [7] the Authors do not specify how to evaluate such numerical relationships while they focus on computational aspects of scaling up with large networks of friends.

3 Two Weighted Social Choice Functions

Generally, group recommendation approaches rely either on building a group profile, resulting from the combination of all the users' profiles, or on merging the recommendation lists, generated for each individual user, using different group decision strategies. We decided to focus on this second approach because it provides us the flexibility required in the group formation process; single user's profiles and recommendations are built independently from the other group's members, and the users' recommendations are merged at the time of providing the group recommendations (e.g., only once the group is established).

Formally, given a group of n users $G = \{1,\ldots,n\}$ and a set of m POI $P = \{1,\ldots,m\}$, each user $i \in G$ has a preference profile $\succ_i$ over P ($\succ_i = \{r_i(1),\ldots,r_i(m)\}$), with $r_i(x) \in \mathscr{R}$, which represents the i's rank for the x POI, and it is, eventually, retrieved by some recommendation mechanism. The goal of the proposed work is to implement social choice functions $SC : \succ^n \rightarrow \succ_G$, that aggregates all the preferences profiles in $\succ_G = \{r_G(1),\ldots,r_G(m)\}$, that is the correspondent ranking as evaluated for the group. In this work, we suppose to have complete and available preference profiles (e.g., they are directly expressed by the users). In future works such preferences will be extracted with the use of recommendation processes.

Differently from typical movie group recommendation, here the problem is not to select a single item for the group, but a subset of activities (top K) for close friends. Hence, there is room for potentially accommodate the needs of each group member. In our framework, a majority evaluation of activities may have the potential risk to leave some users dissatisfied, while, in our opinion, a close group will try to satisfy everyone needs, as long as possible. Finally, according to [8], users involved in real interaction seem to care about fairness and to avoid misery. For these reasons we decided to use a fairness strategy and one based on average satisfaction. Our idea is to weigh such functions with a measure of the influence of each user on the other group's members, and consequently on the group's final decision. To make this, we evaluate the weight of the relationship between pairs of users from the analysis of the interactions on an OSN.

In particular, we are interested in the analysis of the strength and the directionality of online social interactions in order to gather useful information on intra-group relationships, and use the strength of the different pairwise relationships in an aggregated manner in order to evaluate the power/dominance of each single member on the whole group. To compute the users' ranking, we decided to use a simple "non semantic" approach defined in our previous work [9]. For each user $i \in G$ we evaluate his/her dominance value, as the value $D(i) \in [0,1]$. Dominance values are computed by analyzing the popularity of each user within the group, and evaluating the number of directed interactions of each user towards the other group's members. Such popularity values are obtained implementing an extension of the well-known *PageRank* algorithm [10] starting from the users' interactions on the social network *facebook.com*. As in the classic PageRank, each user inherits a portion of popularity from other users.

x	1	2	3	4	5	6	7	8
$r_1(x)$	1	**5**	3	1	2	**$\underline{5}$**	**4**	3
$r_2(x)$	3	**$\underline{4}$**	1	2	**5**	3	2	**4**
$r_3(x)$	1	3	2	**5**	1	4	**3**	**$\underline{2}$**
$r_{fair}(x)$	1	7	3	2	5	8	4	6

Table 1. An example of r_{fair}. Users are ordered from 1 to 3 and $K = 3$. The numbers in bold represent the ratings of the user's K preferred POI, while the rating values corresponding to the POI that causes the least misery are underlined.

3.1 Fairness Strategy

The proposed fairness strategy (r_{fair}) tries to accommodate everyone in the group, and requires an ordering of the users. The first user's top K choices are selected. Among them, the one that causes the least misery to the others is selected (in case of items with the same rating a non deterministic choice is made), and the process is repeated with the successive user in the rank. The group POI ranking values are assigned in a descending order from m to 1. Finally, the group recommendation will correspond to the K POI with the highest r_{fair} values. An example of application of this strategy is provided in Table 1.

One of the main issues with the use of this strategy is that, by changing the users' ordering, the selection process will produce a different result in the outcome, hence, we propose to use the $D(i)$ values to provide a ranking/to sort the users.

3.2 Average Satisfaction Strategy

As a second strategy, we developed an average satisfaction strategy (r_{avg}). Inspired by [1], we defined a strategy that takes into account the dominance as a weight for the rating provided by the user (note that the sum of the dominance values in a group is equal to one). Moreover, a second factor, which can be considered in the evaluation, is a measure of dissimilarity among the users individual ratings. The proposed strategy to evaluate the group G rating for the item x is:

$$r_{avg}(x) = \alpha \cdot \frac{1}{n} \sum_{i \in G} (D(i) \cdot r_i(x)) + \beta \cdot (1 - \sigma^2_{r_i(x)}) \quad (1)$$

where, $r_i(x)$ is the rate for the item x, made by the user i, $D(i)$ is evaluated according to [9], and $\sigma^2_{r_i(x)}$ is the variance of $r_i(x)$ that accounts for the dissimilarity among the ratings of all the $i \in G$ for the item x. α and β are weights. Once that the set $\succ_{avg} = \{r_{avg}(1), \ldots, r_{avg}(m)\}$ is obtained, the final recommendation for the group G will correspond to the first K activities x with the highest $r_{avg}(x)$ values.

4 Experiments and Evaluation

To evaluate the proposed functions, we had to address the problem that, in our specific domain, there are no dataset that can be used for the evaluation. In fact, our application would require a dataset containing information about users interactions on a social network, information on the preferences of individual users, and information on the final choices of the groups, in order to apply the proposed techniques and compare the obtained results. A dataset like this does not exist, so we decided to conduce two pilot studies with real users involved in the task of planning a trip in a city. In the first case, the aim is to evaluate the benefits of the introduction of the dominance values as weights or to order users in the proposed functions. In the second, we focus on the users' satisfaction with respect to the recommendations proposed by the system.

4.1 A User Study with Binary Decisions

In the first case study, we evaluated the behavior of 14 groups composed, in the average, of 3.36 close friends. 46 users took part in the experimentation (26 men and 20 women). The average age was 27.3 with a graduate education.

We asked each user to register on a specific web site using the credentials of *facebook.com*. Once registered, they were asked to imagine to plan a one-day visit in a specific city and to select three activities (from a checklist of ten items) and two restaurants (from a check list of eight) for the day. Since we do not want the users to be involved in strategic reasoning, we did not ask the users to express numerical ratings for the selected choices in this first setting. Hence, we will assign $r_i(x) = 1$ if the user i selects the POI x, and $r_i(x) = 0$ otherwise. The group was, then, asked to discuss, face-to-face, in order to obtain a shared and unique decision for the group. This final decision corresponds to the set $\succ_{GT}$ used to evaluate our functions.

In order to evaluate our results, we adapted the proposed fairness strategy to binary selections. Since a single vote is associated with each POI (0 or 1), at each iteration a user (selected according to his/her dominance value in a descending order) proposes its first K choices (with $K = 3$). For each of the K proposals the votes made by the other users are summed, and the choice with the higher sum is selected. Note that if K is equal to the number of possible outcomes (as in this case), an activity, selected by all the member of the group, will be selected in the final decision. Finally, in order to evaluate the impact of the users' dominance, we also implemented a standard fairness version with a random users ordering ($r_{st.fair}$).

As second strategy, we developed the weighted average satisfaction strategy. With binary choices, the evaluation of dissimilarity does not have a relevant impact on the final decision. Hence, the weights associated to Eq. 1 are $\alpha = 1$ and $\beta = 0$. As in the previous case, we also implemented the standard version of a simple averaging function ($r_{st.avg}$).

We evaluated the similarity of the recommendation provided using the weighted version of the fairness function ($\succ_{fair}$), average satisfaction function ($\succ_{avg}$), and the standard implementation of such functions (i.e., $\succ_{st.fair}$ and $\succ_{st.avg}$) with respect to the groups' ground truth ($\succ_{GT}$).

Table 2. Rate of similairty for $\succ_{st.fair}$, $\succ_{fair}$, $\succ_{st.avg}$ and $\succ_{avg}$ with respect to $\succ_{GT}$

Similarity %	$\succ_{st.fair}$	$\succ_{fair}$	$\succ_{st.avg}$	$\succ_{avg}$
$\succ_{GT}$	61 ± 20	66 ± 18	64 ± 16	74 ± 12

Aggregated results are reported in Table 2. With respect to their standard implementations, functions that take into account social relationships perform slightly better (66% w.r.t 61% for fairness, and 74% w.r.t. 64% for average). A bigger improvement was noted for r_{avg}. On the average, the r_{avg} social choice function performs better than the r_{fair}, often guessing 4 on 5 activities. In many cases the responses of both strategies were the same, and there is only one case in which r_{fair} performs better than r_{avg}. Fairness strategy, in this simple binary case, with no information about rankings, suffers more of random choices made by the function in the case of activities with the same final score.

4.2 A User Study with Rankings

In this second case study, we evaluated the behavior of 17 new groups of friends composed, in the average, of 3.1 people. 53 users took part in the experimentation (26 men and 27 women). The average age was 26.8 with a graduate education.

This experiment was divided into two phases. In the first, as in the previous user study, each person was asked to register on a specific web site using the credentials of *facebook.com*. Once registered, he/she was asked to imagine planning a one-day visit in a specific city, but this time also to provide ratings (from one to five stars) to the ten proposed activities for the day and to the eight restaurants. Each rate corresponds to how likely it would be for the user to visit such place (or to eat in). Once that all the member of a group completed this first phase, we separately invited each member to login again and to complete the process by evaluating the proposed recommendations.

In the r_{fair} strategy the users were selected according to their dominance values in a descending order, with $K = 3$. In the r_{avg} strategy the weights associated to Equation 1 were $w_1 = 0.8$ and $w_2 = 0.2$ as suggested in [1].

Users were presented with both the recommendation provided by using the r_{fair} and r_{avg} functions (as in the previous case the top five activities were recommended). Moreover, the associated ratings for the proposed activities provided by all the members of the group in the first phase were shown.

Finally, each user was requested to answer the following questions:

1. Which of the two proposed itineraries do you prefer? [None/First/Second/Both]
2. How do you rate the first itinerary? [1 to 5]
3. How do you rate the second itinerary? [1 to 5]
4. How much have you been influenced in the evaluations by the knowledge of your friends' ratings? [1 to 5]

Table 3. Aggregate results for the second user study

Accept. %	$\succ_{fair}$	$\succ_{avg}$	$\succ_{fair} + \succ_{avg}$
average	66 ± 32	62 ± 30	81 ± 20

(a) Acceptance rate for the $\succ_{fair}$, $\succ_{avg}$ and globally $\succ_{fair} + \succ_{avg}$.

Users' rates	$\succ_{fair}$	$\succ_{avg}$	*friends*
average	4.3 ± 0.6	4.2 ± 0.5	2.6 ± 0.6

(b) Average ratings for the $\succ_{fair}$, $\succ_{avg}$ and for the evaluation of the friends' influence.

In Table 3a, we reported the acceptance rate mean percentage. In the average, two out of three members of each group accepted the proposed solution evaluated by the r_{fair}, while for r_{avg} average value is a little bit smaller. Considering both the proposed options the system had an acceptance rate of 81%. Moreover, considering directly all the users involved in the test, and not averaging on groups, 49 people out of 53 (i.e., 93% of the users) accepted at least one of the proposed itinerary.

In Table 3b, we reported the mean values for the users' ratings of the $\succ_{fair}$ proposed itinerary (i.e., question number 2), and $\succ_{avg}$ proposed itinerary (i.e., question number 3). Both the proposed results got, on average, a good appreciation by the users (more than 4 stars). Moreover, we evaluated the Pearson correlation index ρ between such ratings and the acceptance rates of the proposed strategies. As we expected, there is a strong correlation between the acceptance rate of the proposed strategies and their evaluations made by the users. The obtained value for $\succ_{fair}$ is $\rho = 0.67$ (with a significance $p = 0.003$), while for $\succ_{avg}$ is $\rho = 0.77$ (with $p = 0.0003$).

Surprisingly, the average evaluation of the impact of the friends' opinions on the evaluation of results ($friends$ in Table 3b) is 2.6. We imagine that such value is an effect of the testing procedure that ends with a private evaluation of the proposed solutions for the group.

5 Conclusion

In this paper, we show that a simple, but effective, approach to extract social relationships can be used as a weight or to order users in the definition of social choice functions. With respect to their standard implementation, the weighted functions had better performances. In particular, in the first user study with binary selections, a bigger improvement was noted for the average satisfaction function, which also performed better than the fairness that suffers more of random choices. On the contrary, in the second one, which involved POI ranking, fairness strategy has a bigger acceptance rate and appreciation evaluation.

Providing a solution for the whole group is not the last step in group recommendation since, differently from the case of one item recommendation (a movie to watch or a restaurant), when the recommendation involves a set of activities, the group may be involved in a subsequent interaction in order to take the final decision. Of course, the better is the recommendation, the easier is for the group to reach a consensus. In this direction, our next step will be to provide the group with interfaces to reach this final decision.

Acknowledgments. The research leading to these results has received funding from the Italian Ministry of University and Research and EU under the PON OR.C.HE.S.T.R.A. project (ORganization of Cultural HEritage for Smart Tourism and Real-time Accessibility).

References

1. Gartrell, M., Xing, X., Lv, Q., Beach, A., Han, R., Mishra, S., Seada, K.: Enhancing group recommendation by incorporating social relationship interactions. In: Proc. of the 16th ACM International Conference on Supporting Group Work, pp. 97–106. ACM (2010)
2. Cantador, I., Castells, P.: Group recommender systems: New perspectives in the social web. In: Recommender Systems for the Social Web. Intelligent Systems Reference Library, vol. 32, pp. 139–157. Springer (2012)
3. Senot, C., Kostadinov, D., Bouzid, M., Picault, J., Aghasaryan, A., Bernier, C.: Analysis of strategies for building group profiles. In: De Bra, P., Kobsa, A., Chin, D. (eds.) UMAP 2010. LNCS, vol. 6075, pp. 40–51. Springer, Heidelberg (2010)
4. O'Connor, M., Cosley, D., Konstan, J.A., Riedl, J.: Polylens: A recommender system for groups of users. In: Proc. of the 7th European Conf. on CSCW, pp. 199–218 (2001)
5. Garcia, I., Sebastia, L.: A negotiation framework for heterogeneous group recommendation. Expert Systems with Applications 41(4), 1245–1261 (2014)
6. Carvalho, L.A.M.C., Macedo, H.T.: Users' satisfaction in recommendation systems for groups: An approach based on noncooperative games. In: Proc. of the 22nd Int. Conf. on World Wide Web Companion, WWW 2013, pp. 951–958 (2013)
7. Salehi-Abari, A., Boutilier, C.: Empathetic social choice on social networks. In: 13th International Conference on Autonomous Agents and Multiagent Systems, pp. 693–700 (2014)
8. Masthoff, J.: Group recommender systems: Combining individual models. In: Recommender Systems Handbook, pp. 677–702. Springer US (2011)
9. Caso, A., Rossi, S.: Users ranking in online social networks to support pois selection in small groups. In: Extended Proceedings of the 22nd Conference on User Modelling, Adaptation and Peronalization, UMAP 2014, pp. 5–8 (2014)
10. Brin, S., Page, L.: The anatomy of a large-scale hypertextual web search engine. Comput. Netw. ISDN Syst. 30(1-7), 107–117 (1998)

Designing a User Interest Ontology-Driven Social Recommender System: Application for Tunisian Tourism

Mohamed Frikha, Mohamed Mhiri, and Faiez Gargouri

University of Sfax - MIRACL Laboratory - Sfax, Tunisia
{med.frikha,med.mhiri}@gmail.com,
faiez.gargouri@isimsf.rnu.tn

Abstract. The tremendous growth of online social networks all over the world has created a new place and means of social interaction and communication among people. This paper aims to improve traditional recommender systems by incorporating information in social networks, including user preferences and influences from social friends. A *user interest ontology* is developed to make personalized recommendations out of such information. In this paper, we present a *social recommender system* employing a *user interest ontology*. Our system can improve the quality of recommendation for Tunisian tourism domain. Finally, our *social recommendation algorithm* will be implemented in order to be used in a Tunisia tourism Website to assist users interested in visiting Tunisian places.

Keywords: Recommender Systems, Collaborative Filtering, Social Network, User Interest Model, Social Semantic Web, Ontology.

1 Introduction

Since social networks are developed by several different kinds of relationships, it remains impossible for the graphs' edges and the numerical values to explain all the semantic relationships. Having mentioned this problem, we propose, as a solution, a method that makes it possible to represent a social network based on ontology. Using an ontology-based method could allow us to describe all the semantic relationships and the interactions in a social network. The importance of the use of ontologies to represent the types of actors and the relationships in a social network for the purpose of semantically visualizing the databases has been demonstrated by [1]. Based on the fact that ontologies can be used to semantically describe social relationships and interactions and that social networks can generate and suggest several varied recommendations with reference to the user needs [2], this paper aims to exploit the impact of online social network (Facebook) on the consumers' purchasing decision process in Tunisian Traveling Websites. To achieve this purpose, we have developed a user interest ontology that represents all the preferences and interests of a user. Those preferences and interests are extracted from the user's profile in the social network.

Based on mainstream literature on recommender systems in social networks [3], social networks already propose to their users recommendations based on their profiles. In a similar fashion, they help users find people for sharing common social

J. Bajo et al. (eds.), *Trends in Prac. Appl. of Agents, Multi-Agent Sys. and Sustainability*,
Advances in Intelligent Systems and Computing 372, DOI: 10.1007/978-3-319-19629-9_18

activities and preferences. Being aware of the importance of the user and of his/her preferences in recommendations, we study, in this paper, the impact of incorporating semantic user profile (derived from past users' behaviors and preferences) on the accuracy of a recommender system[4],[5].

The rest of the paper is organized as follows. Section 2 describes the need of a social network-based recommender system, while Section 3 shows and accounts for the use of a personalized social recommender system for Tunisian tourism information; a new paradigm of tourism recommender systems developed by using information in social networks. Section 4 presents the need for a semantic user profile in the social network and explains our method for modelling a user interest ontology to personalize social recommender system. Finally, section 5 concludes the paper and describes the orientations of our future work.

2 Social Network-Based Recommender Systems

Traditional recommender systems suffer from many issues. For example, in order to measure item similarity, content-based methods rely on explicit item descriptions. However, such descriptions may be difficult to obtain for items like ideas or opinions. Collaborative filtering has the data sparsity problem and the cold-start problem [6]. In contrast to the huge number of items in recommender systems, each user normally only rates a few. Therefore, the user/item rating matrix is typically very sparse. It is difficult for recommender systems to accurately measure user similarities from those limited number of reviews. A related problem is the cold-start problem. Even for a system that is not particularly sparse, when a user initially joins, the system has none or perhaps only a few reviews from this user. Therefore, the system cannot accurately interpret this user's preference.

Integrating social networks in recommender systems can, therefore, result in more accurate recommendations. That is to say, the information, interests, and recommendations retrieved from social networks can improve the prediction accuracy. Additionally, the information obtained about the users and their friends makes it unnecessary to look for similar users and to measure their rating similarity as the fact that two people are already friends can imply that they have things in common [7]. The data sparsity problem can be solved in this case. The cold start issue can also be overcome because even if a user has no interests' history, recommendations can still be made based on a friend's preferences.

All of these intuitions and observations motivate us to adopt this approach of social network-based recommender system that can take advantage of information in social networks. Our goal is to represent all these information extracted from social network in a *user interest ontology* [8] that can help us in the recommendation process. In fact, in order to be able to recommend an object to the user, one must know the interests that represent a part of the user profile.

3 Personalized Social Recommender System for Tunisian Tourism Information

3.1 Framework of Personalized Social Recommender System

Our primary objective in this study is to apply a personalized recommender system to tourism information service. The reason behind this application is to provide tourists with better information searching experiences and to improve the individuation of tourism information. In a personalized social recommender system, the user connects to his/her social network, and then the system extracts the preferences and relations to determine the user's interest as well as his/her friends'. After that, the system generates specific recommendations by employing a recommendation algorithm which leads to the presentation of the results.

When the user connects to the social network (Facebook), the system collects all the data from the user's profile and extracts his/her interests. Those user interests are presented in the form of an ontology. Our social recommender system, however, needs the user interest's ontology to calculate similar items to the user's preferences. These items will be recommended to the user. Figure 1 further illustrates the functioning of our framework.

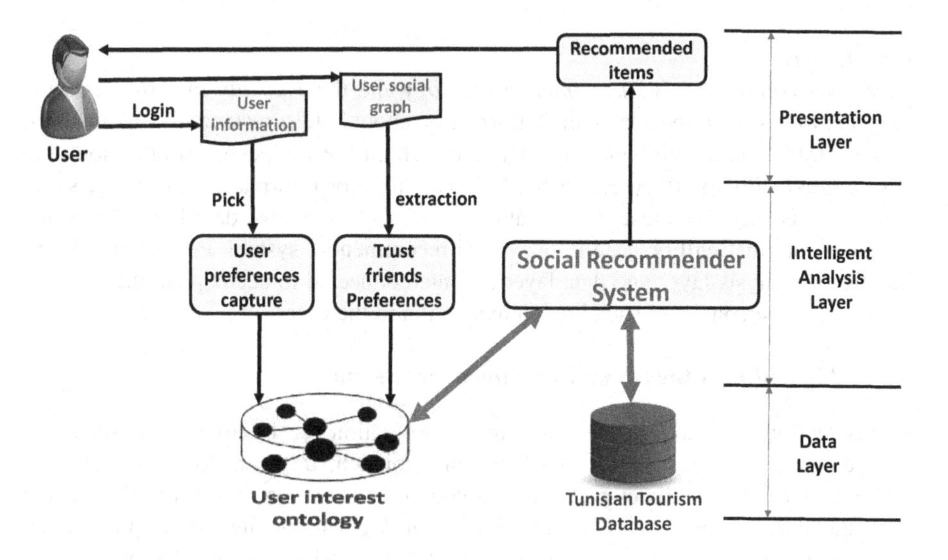

Fig. 1. Framework of social tourism recommender system based on user interest ontology

Presentation Layer

The purpose behind a presentation layer is to allow the interaction between the user and the system and to present the recommended items to the user. This layer picks the user's profile when the user connects to his/her account, gathers data and displays information through websites. It acts as the interface for the user to access the system.

By using general websites as interfaces, it enables the user to visit items which are recommended to him/her by the system. The presentation layer mainly consists of Graphical User Interface and the results of the recommendation system.

Intelligent Analysis Layer

Intelligent analysis layer is made up of intelligent analysis module and the social recommendation system. Intelligent analysis module is composed of two modules of user preference capture and friends' preference capture. User preference is collected from user's profile, and friends' preference is collected from the user social graph presented by Facebook. Before capturing the user preference, we must first analyze the entire user profile, the personalized friends' list; the News Feed; the relations and the "likes" of the users and their friends. All these preferences will be represented in a *user interest ontology* (detailed in the next section). Then the social recommendation system generates recommendations with the help of the *user interest ontology*. By using social recommendation algorithm, our system can find similar items to this user's interest ontology in the *Tunisian tourism database*. Our intelligent recommendation system picks out information in agreement with the user's needs from the *Tunisian tourism database* and produces recommendation set. In this layer, the system can offer a dynamic guidance for the user's choice when he/she desires to travel to Tunisia.

Data Layer

Data layer consists of the *user interest ontology* and the *Tunisian tourism database*. *Tunisian tourism database* contains information about hotels, restaurants, monuments, musses, travels and entertainments. The aim behind these types of information is to ensure satisfaction on the users' part when searching for information about tourism in Tunisia. It is the storehouse for operation data. In other words, data layer stores and manages data that will be used in our social recommender system. Presentation layer, intelligent analysis layer and data layer are interconnected to accomplish the function of providing personalized social recommendation to the user.

3.2 Algorithm of the Social Recommender System

In this section, we show how the semantic information can be used in the social semantic recommender system. Based on our approach, the semantic information is broken down into two major parts: the ontology representation of user interest and the semantic similarity between the *user interest ontology* and the item description in the *Tunisian tourism database*. We, then, present our algorithm of social semantic recommendation to reduce the list of Tunisian tourism items that will be recommended to the user; the most similar items to the *user interest ontology* will be recommended and rated by the user for the purpose of updating our *user interest ontology*. The recommendation algorithm is shown in Table 1. We can define the algorithm input as all the items in the *Tunisian tourism database*, and the algorithm output as a list of most items similar to the *user interest ontology*.

Table 1. Steps of the recommendation algorithm

```
1. User connects with his/her Facebook account in our Tunisia
Tourism Website
2. Picks up user information and friends' information
3. Creates the User Interest Ontology
4. Starts the search in our Tunisia Tourism Database
5. For each item in the Database
       (a) Compare the user interest ontology with the descrip-
tion of this item (Similarity)
       (b) Compare the similarity degree between user interest
and item with a threshold.
       (c) If the similarity degree > threshold
                    i. add item to the proposed list
       (d) Else Go to (5)
6. Sort out the proposed list
7. Recommend the most similar item in the list to the user
8. Ask the User to rate the recommended item
9. If User Rating < 2
       (a) Recommend the next item in the proposed list
       (b) Go to (8)
10. Else Update user interest ontology with the proposed item
11. If User wants another recommendation
       (a) Go to (5)
12. End.
```

Our proposed algorithm uses semantic information described in a *user interest ontology*. This ontology contains social information represented in the social network as heuristics integrated in our recommendation algorithm to find out the item most interesting to this user in *Tunisian tourism database*. When a user has access to the Tunisian Tourism Website and connects with his/her account, the system will automatically gather the user information and the friends' comments and likes to create the *user interest ontology*. After that, the system starts the search for similar items to this ontology that represent the preference of the user in our *Tunisian tourism database*. The system will prepare a proposed item list and will sort out items in the list. An item is similar to the *user interest ontology* if it is superior to a prefixed threshold. The top ranked items will be, then, recommended to the user. If the user does not like the recommended item (User Rating < 2), the system will recommend the next item in the proposed list and will ask him/her to rate it. Furthermore, if the user likes the recommended item, our system will update the *user interest ontology* and will search again in the database taking into consideration this update.

4 User Interest Ontology to Personalize Social Recommender System

Our method for extracting user interests consists of several steps. First, we collect user data from his/her profile and behaviors in the network. Then, we analyze the user data to extract the interests of the user. Finally, we present these data in the form of ontology.

4.1 User Data Collection

The user is identified in a social network by his/her traits and behaviors, and his/her interactions with the various system services. In our work, we treat the textual form of data, which presents the major form, to identify user interests. In this context, the content generated by the user can be regarded as explicit and implicit data. First, the explicit data are the data about users which are explicitly declared and which represent user information and preferences. The user information may be either static or dynamic. Static user information are mainly general information such as name, first name, sex, date of birth, etc. while dynamic user information relate to marital status, level of education, place of residence, occupation, email address, phone number, nickname, photo, etc. Then the implicit data are usually discovered through actions provided by the user or by its behavior such as comments, quotes “like", publications, events, etc. and also deduced from explicit data contained in the ontology.

4.2 Data Analyzing and User Interest Extraction

In this step, we analyze the preferences and behaviors of the user (shares, comments, events, etc...). Thus, in this phase, we try to filter the raw data and display them in a representation in order to further process them. Indeed, these textual data are not structured and written in a formal language, which makes their use very difficult in their raw state.

User Preferences Analyzing

To analyze the preferences of the user, which are presented by a set of XML files extracted from the social network. These XML files are obtained with the Facebook API Open Graph that gives us the authorization to accede to the user profile. First, we begin by removing all empty concepts. Then, we construct a matrix of occurrence that can indicate the number of occurrences of each concept. We afterwards, determine the semantic relations between the concepts using "*WordNet*" which covers most common English words, and a *base of terms* containing the common concepts that do not belong to "*WordNet*". We use a calculating similarity between concepts when the concept does not belong to "*WordNet*" or to the *base of terms.*

After extracting the various concepts used by the user, we will try to determine the semantic relationships that exist between these concepts. To do this, we use the language tool "*WordNet*" to determine if there are relationships between the concepts of the matrix. In many cases, we can find concepts that do not exist in "*WordNet*". To remedy this, we use a *base of terms* that contains commonly used concepts and acronyms in the tourism domain. When the concept does not exist in the *base of terms*, we use a function that allows the calculation of similarity to check if the concept is similar to another concept in the matrix or in the *base of terms*.

Techniques based on strings are not sufficient when the concepts are semantically close and when their names are different. The query language resource such as "*WordNet*" may indicate that the concepts are similar. In our work, we use the proposed function [9] to calculate the linguistic similarity between two concepts. For the calculation of this similarity, the Syn (c) function calculates the Synsets of the concept C in "*WordNet*". When S = Syn (c1) ∩ Syn(c2) is the set of common sense between the to-be-compared c1 and c2. The cardinality of S is then:

$$\lambda(S) = |Syn(c1) \cap Syn(c2)| \tag{1}$$

When min (| Syn (c1) |, | Syn (c2)|) is the minimum between the cardinalities of the two sets Syn (c1) and Syn (c2). The similarity between the two concepts is then c1 and c2 and is defined as follows:

$$Sim_{ling}(c1,c2) = \lambda(S) / \min(|Syn\,c(1)|, |Syn\,c(2)|) \tag{2}$$

This measure returns 1 if at least c1 is the only synonym to c2 and c2 is the only synonym to c1.

User Behavior Analyzing

The user provides a set of actions that enrich our knowledge about these interests. These actions can be comments (on photos, videos, articles, etc.), publications (such as photo or video sharing), and events of various interests or "I like" signs on the pages or groups. However, we begin by extracting sentences of each document (comments, status, etc.). Then, we apply the same method in the user preference analysis to extract concepts and relations between them. In this step, we use the TF-IDF method [10] to measure the weight (density) of each concept in a document. The TF-IDF algorithm is based on term weighting and has the advantage of being easily used with statistics and generic techniques. In our approach, to obtain the interests of the user, we remove from the matrix constructed in the previous step the concepts with a lower number of occurrences of a predetermined threshold. For this, we only get the most interesting concepts for the user.

4.3 Determination of Trusted Friends List from the User Social Graph

To determine the list of friends near a user, we begin, first of all, by all the friends who have made interactions or shared information with the user in a very specific period of time. Thus, interactions are of different types. They can be comments or mentions "Like" on objects in the profile (from the user or from friends). Then we calculate the number of occurrences of each friend in these interactions. Finally, we choose friends who have a number of occurrences beyond a predetermined threshold. However, the extraction of the trusted friends list of a user is performed to determine the preferences and interests of each friend. Indeed, these interests can be useful to know the interests of our user.

4.4 Concept Weight in the User Interest Ontology

A key factor in this ontology is the representation of semantic relations obtained in the step of determining the types of relationships between concepts. Each concept in our ontology has a degree of interest (weight) to express the importance of this concept for the user. This weight is calculated using the number of occurrences and the last occurrence time of the concept. This weight represents the degree of interest of the concept to the user.

5 Conclusion and Future Work

In this work, we have presented the role of social networks as sources for the development of recommendation systems. In addition, we proposed a method that personalized social network-based recommender system with *user interest ontology*. Finally, we integrate the *user interest ontology* in a social semantic recommender system to deal with the lack of semantic information in personalized recommendation system in tourism domain. Now, we search to finalize our prototype and make it available online to improve the tourism sector in our country and to assist users interested in visiting Tunisian places. In our future work, we search to evaluate our system and to test the users' satisfaction towards the recommendations proposed by the system.

References

1. Correa, C.D., Ma, K.L.: Visualizing social networks. In: Aggarwal, C.C. (ed.) Social Network Data Analytics, 1st edn., pp. 307–326. Springer (2011)
2. Middleton, S.E., Roure, D.C.D., Shadbolt, N.R.: Ontology-based Recommender Systems. In: Staab, S., Studer, R. (eds.) Handbook on Ontologies, 2nd edn. Springer (2009)
3. Zhou, X., Xu, Y., Li, Y., Josang, A., Cox, L.: The state-of-the-art in personalized recommender systems for social networking. Artificial Intelligence Review 37(2), 119–132 (2012)
4. Sieg, A., Mobasher, B., Burke, R.: Improving the effectiveness of collaborative recommendation with ontology-based user profiles. In: Proc. of Intl. WIHFR, pp. 39–46 (2010)
5. Su, Z., Yan, J., Chen, H., Zhang, J.: Improving the preformance of personalized recommendation with ontological user interest model. In: Seventh International Conference on Computational Intelligence and Security (2011)
6. Adomavicius, G., Tuzhilin, A.: Toward the next generation of recommender systems: A survey of the state-of-the-art and possible extensions. IEEE Transactions on Knowledge and Data Engineering 17(6), 734–749 (2005)
7. He, J., Chu, W.W.: A Social Network-Based Recommender System (SNRS). In: Memon, N., Xu, J.J., Hicks, D.L., Chen, H. (eds.) Data Mining for Social Network Data. Annals of Information Systems, vol. 12, pp. 47–74 (2010)
8. Frikha, M., Mhiri, M., Gargouri, F.: Toward a User Interest Ontology to Improve Social Network-Based Recommender System. In: Sobecki, J., Boonjing, V., Chittayasothorn, S. (eds.) Advanced Approaches to Intelligent Information and Database Systems. SCI, vol. 551, pp. 255–264. Springer, Heidelberg (2014), doi:10.1007/978-3-319-05503-9_25
9. Fellah, A., Malki, M., Zahaf, A.: Alignement des ontologies: utilisation de WordNet et une nouvelle mesure structurelle. In: Conférence en Recherche d'Information et Applications, CORIA, pp. 401–408 (2008)
10. Salton, G., Wong, A., Yang, C.S.: A Vector Space Model for Automatic Indexing. Communications of the ACM 18(11), 613–620 (1975)

Part V

Special Session on Agent-Based Modeling of Sustainable Behavior and Green Economies (AMSBGE)

Designing Decision Trees for Representing Sustainable Behaviours in Agents

N. Sánchez-Maroño[1], A. Alonso-Betanzos[1], O. Fontenla-Romero[1],
J.G. Polhill[2], and T. Craig[2]

[1] University of A Coruña, A Coruña, 15071, Spain
{noelia.sanchez,amparo.alonso.betanzos,oscar.fontenla}@udc.es
[2] The James Hutton Institute, Craigiebuckler, Aberdeen. AB15 8QH
{gary.polhill,tony.craig}@hutton.ac.uk

Abstract. Decisions made by workers in their daily routine have an environmental impact. The LOCAW project has analyzed the drivers and barriers for an employee to choose a particular option in large organizations. In this project, Agent-Based Models (ABM) seek to clarify interactions among relevant actors and provide insights into the necessary conditions to achieve more sustainable organizations. For theoretical and practical reasons, it was considered to use decision trees to represent the internal behavior of the agents in the model. This paper focuses on how to improve the generalization capabilities of these decision trees using feature selection and discretization techniques. The application of these techniques is intended to obtain simpler decision trees, but more accurate. Experimental results of three daily activities support the adequacy of the approach presented.

Keywords: Decision trees, feature selection, discretization, agent-based modeling.

1 Introduction

In addition to deduction and induction, simulation is sometimes seen as a third methodology for doing research. Even though simulation does not prove theorems, it can enhance our understanding of complex phenomena that have been out of reach for deductive theory [1]. Agent-based modelling and simulation (ABMS) is a relatively new approach to modelling systems composed of autonomous, interacting agents. Agent-based modelling is a way to model the dynamics of complex systems and complex adaptive systems. Agent-based models also include models of behaviour (human or otherwise) and are used to observe the collective effects of agent behaviours and interactions [2].

Using these kind of techniques, the LOCAW project (LOw Carbon At Work, `http://www.locaw-fp7.com`) has analysed (un)sustainable behaviour and practices in the workplace and it has focused on the factors determining these actions in six large organizations in the public as well as the private sector. The conclusions of this analysis have then been used as input for the design of agent-based models. The ABM was addressed by considering two submodels: a) the decision-making model, responsible for representing the internal behavior of agents, i.e. how they choose an option in their daily routines, and b) the social network that reflects interactions between

J. Bajo et al. (eds.), *Trends in Prac. Appl. of Agents, Multi-Agent Sys. and Sustainability*,
Advances in Intelligent Systems and Computing 372, DOI: 10.1007/978-3-319-19629-9_19

agents. Two restrictions were considered when designing the decision-making model, first, it has to be based on actual data and, second, experts in the domain have to be able to analyze and discuss it. For that reason, decision trees were selected.Then, the ABM represents the daily routine of employees in an organization, where they have to choose alternatives to different actions. The decision trees determine the choice of an agent based on its own internal parameters (gender, position, values, norms...), while the social network may change some of these internal parameters by influence of other agents. Notice that an agent's option affects other agents which in turn influence the agent itself [3].

According to [4], agents follow simple rules. We, as humans, follow rules, in the form of norms, conventions, protocols, moral and social habits and heuristics. Although the individual rules may be quite simple, they can produce global patterns that may not be at all obvious and can be very difficult to understand. Thus, decision trees derived from data should not be necessarily complex, as this fact could prevent expert analysis. Therefore, if the decision-making is based on reduced set of features, it is possible to execute different simulations to analyze the effects obtained when varying them. As the main goal of LOCAW project is to reach more sustainable organizations, it is important to identify those features to promote the adequate interventions in the organization.

In this paper, we focus on how to improve the decision-making process, using decision trees, with good generalization capabilities based on a reduced set of features, enough to generate variability in the model, but without hindering the work of experts to analyze them. For achieving this goal, feature selection and discretization techniques were used before applying a classical decision tree algorithm. The results achieved for the University of A Coruña (UDC), one of the public organizations studied by LOCAW consortium, will be presented.

2 Materials: Actual Data from a Questionnaire

Different quantitative and qualitative tools were applied in LOCAW project to obtain relevant information from the different organizations to model. Among them, a questionnaire was designed to get actual data from the personnel and with the aim of examining relationships between individual factors and pro-environmental behaviour at work. The questionnaire is divided into three parts. First, there are six general questions about the personal situation of the employees (such as age and gender) and the extent to wich they believe to have an exemplary role in their organization. This was followed by the second part comprising 61 questions about motivational factors (i.e., values and environmental self-identity). To facilitate the design of the social network of the agents in the ABM, at the end of this section, a question about the number of personal relationships was included. Third, participants completed a set of questions on pro-environmental behaviour at work. In order to fulfil the objectives of LOCAW project, the questions in this section were focused on three main categories of practices:

- Organization-related mobility
- Energy and materials consumption
- Management and generation of waste.

The questionnaire could be anonymously fulfilled by employees using a web platform. However, the number of respondents, as expected, was very low. In fact, for the University of A Coruña, an organization with more than 2000 employees, around 300 responses were obtained, many of them partially fulfilled, so they have to be discarded. After analyzing all of them, only 197 were retained. Despite this low level of participation, it was the organization with the highest number of responses among the six organizations included in LOCAW and, thus, selected for this study.

3 Obtaining Decision Trees with Good Generalization Capabilities

As mentioned in the introduction, the decision-making process has to be both data-driven and theoretically-consistent. For that reason, decision trees were elected to determine each of around 40 different work behaviours. For the sake of brevity, we will focus on three of them, one for each topic covered by LOCAW project:

1. Mobility: *When you commute, how often do you commute by car?*
2. Energy: *How often do you have the lights on at your workspace when there is no one in there?*
3. Management of waste: *How often do you separate your plastic from the regular garbage at work?*

A decision tree has been generated for each one of these behaviours taking as inputs the multivalued answers to the first two parts of the questionnaire (gender, values, etc.). Each desired output has 7 possible values because they are items from a Likert-scale (range from $1-$*Never* to $7-$*Always*). Our first attempt was to apply the well-known *C*4.5 algorithm [5] over the available data (197 samples) that were randomly split into training set and test set (2/3-1/3). Although the accuracy results were very good for the training set, they turned to obtain very poor performance results in the test set with accuracy values under 40%. These results clearly showed that the trees are overfitted, i.e, with poor generalization capabilities, although pruning techniques were used. Moreover, a look at the decision trees generated indicates that they are not using all available inputs (68), but a rather smaller number (around $20-24$, depending on the behaviour). Despite this feature selection inherently done by the *C*4.5 algorithm, the complexity of the decision trees is still high with more than 30 leaves in some cases. Therefore, we decide on using feature selection techniques to reduce the input dimensionality and, consequently, the complexity of the trees, improving their generalization capabilities. It was also noted that not all possible outcomes were covered by the available sample. For example, in the plastic separation behaviour most of the answers are at the ends (1 or 7), but there is almost no representation of the intermediate points $(3-4)$. For that reason, a discretization step was also considered of interest. Both preprocessing methods, discretization and feature selection, are briefly described in the following subsections. All of them were applied using Weka tool [6].

3.1 Feature Selection

Feature selection (FS) is the process of detecting the relevant features and discarding the irrelevant ones for a given data set. A correct selection of the features can lead to an improvement of the inductive learner, either in terms of learning speed, generalization capacity or simplicity of the induced model [7]. Feature selection methods can be divided into three models: filters, embedded and wrapper methods, depending on the degree of dependence with the subsequent classifier, so filters are completely independent, whereas wrappers use the classifier performance to select the features. As the number of feature selection methods is constantly growing, we opted for classical methods covering the three existing types. Then, Correlation-based Feature Selection (CFS) was selected as filter [8] whereas Support Vector Machine-Recursive Feature Elimination (SVM-RFE) as embedded [9]. In the wrapper approach, the feature selection algorithm conducts a search for a good subset of features using the classifier itself as part of the evaluation function. Then, it depends on both the classifier and the search strategy. It is expected to obtain better results when the same classifier is used in the selection and in the subsequent classification. For that reason, C4.5 was used as induction algorithm. However, to test this issue and for the sake of completeness, the naive-Bayes (NB) classifier was also considered. Regarding the search strategy, two opposing alternatives were chosen: backward and forward selection. Then, four different wrappers were tested: C4.5-BACK, C4.5-FOR, NB-BACK and NB-FOR. In summary, six different methods are tested, all of them based on different metrics. Thus, completely different subset of features may be obtained [7].

It should be noted that all these methods return a subset of features except SVM-RFE which returns a ranking of features. Therefore, a threshold is needed, in this case two values were chosen: 20 and 30 top features in the ranking (SVM-RFE20 and SVM-RFE30), values close to those used by C4.5 algorithm when applied alone. It is important to remark that other well-known FS methods were used (INTERACT, Information Gain or Consistency), however, the small number of features returned (even 0), forced us to discard them.

3.2 Discretization

The studied behaviors have 7 possible output values in a Linkert scale, as detailed above. However, since the number of samples is reduced, not all of them are represented. Therefore, the desired output was discretized. As the attribute to be discretized is the class itself, the algorithm must be unsupervised. Then, a binning technique was used which is the simplest method to discretize a continuous-valued attribute by creating a specified number of bins. The bins can be created by equal-width (EWD) and equal-frequency (EFD) [10]. In equal-width, the continuous range of a feature is evenly divided into intervals that have an equal-width and each interval represents a bin. In equal-frequency, an equal number of continuous values are placed in each bin. The two methods are very simple but are sensitive to the number of bins, therefore, two different number of bins were tested, maintaining the symmetry of the Likert scale (5 and 3). Proportional k-interval discretization [11], a well-known and used method, was also applied. However, this method automatically determines the number of bins and it returned 7 bins for each behavior evaluated, thus preventing its use in our problem, as the effect pursued is not achieved.

4 Experimental Results

Our aim is to be able to obtain decision trees that exhibit good generalization results. To achieve this, the procedure was carried out as follows:

1. A method of feature selection was applied on the complete data set.
2. The output class is discretized by binning, testing 3 and 5 bins.
3. The data set is randomly split (2/3 for training and 1/3 for testing).
4. The decision tree is built by applying the C4.5 algorithm to the training set.
5. This learnt tree is later applied to the test set to obtain the accuracy.

As the evaluation may depend heavily on which data points end up in the training set and which end up in the test set, the procedure (from step 3 to 5) was repeated 5 times to obtain more reliable results. These mean results are shown in tables 1-3, for mobility, energy and waste related behaviours, respectively. The results obtained by applying only one or none of the two preprocessing steps are also shown in the tables denoted as no discretization (No-Disc) or no feature selection (No-FS).

As stated at section 3, the data set is clearly unbalanced. The considered behaviours denote that most of the university employees commute by car (70,6%) and the same proportion of employees turn the lights off when going home. Related to plastic separation, two opposite options are usually elected, 32,49% of employees never separate plastic and 25,38% always do it. Then, for the first two behaviours (mobility and energy) is possible to achieve a good level of accuracy without any classification. This was the case for most of the decision trees with lower number of pre-selected features (CFS, NB-FOR, and C4.5-FOR rows in tables 1-2) where all instances are allocated to the majority class. Regarding the same rows in table 3, the trees generated distinguish between the two extreme outputs (1 or 7), however, they are based on very few features. None of these results are useful for the purpose of designing the agents in the model. Then, to provide relevant results, we have focused on those that may have lower accuracy, but do distinguish several classes and therefore, they allow for introducing variability in the model (i.e., rows NB-Back, C4.5-Back and both alternatives of SVM-RFE). In all these rows and No-FS row, it can be considered that EWD with 3 bins exhibit very good performance result (best mean results in 10 of 12). Therefore, to summarize, Table 4 presents relevant information about the decision tree that achieves the best accuracy (from the set of 5 run) for each combination with EWD bining in 3 intervals. These experimental results clearly show that the methods employed improve accuracy in the test set, while reducing the complexity of the decision tree. Note that to overcome the unbalanced problem, the Synthetic Minority Over-sampling Technique (SMOTE) [12] was applied to increase the number of samples of the minority class, however, being a multiclass problem, the experimental results do not improve those in Table 4.

Table 1. C4.5 Mean Accuracy Results for Mobility related Behavior. Number of features selected by each FS method in parenthesis.

FS Method	Acc	Discretization Method				
		No-Disc	5 bins		3 bins	
			EWD	EFD	EWD	EFD
No-FS (68)	Train	87,33	91,71	93,58	93,74	94,68
	Test	36,76	55,15	45,81	55,15	49,19
CFS (8)	Train	60,73	71,37	60,73	71,37	63,23
	Test	64,51	74,49	64,51	74,49	61,39
NB-BACK (32)	Train	84,82	87,49	91,40	87,33	94,68
	Test	44,84	60,16	42,68	62,62	49,81
NB-FOR (1)	Train	60,73	71,37	60,73	71,37	61,98
	Test	64,51	74,49	64,51	74,49	58,89
C4.5-BACK(46)	Train	84,82	89,05	91,87	91,55	95,15
	Test	40,19	54,20	43,61	57,96	48,57
C4.5-FOR(4)	Train	60,73	71,37	63,85	71,37	61,98
	Test	64,51	74,49	60,76	74,49	58,89
SVM-RFE(20)	Train	77,31	76,39	87,33	85,43	91,25
	Test	48,60	69,21	47,68	60,11	51,42
SVM-RFE(30)	Train	76,85	83,72	89,98	89,36	92,65
	Test	48,32	62,01	43,94	53,27	53,91

Table 2. C4.5 Mean Accuracy Results for Energy related Behavior. Number of features selected by each FS method in parenthesis.

FS Method	Acc	Discretization Method				
		No-Disc	5 bins		3 bins	
			EWD	EFD	EWD	EFD
No-FS (68)	Train	85,87	91,23	89,59	95,30	93,67
	Test	34,79	59,41	39,09	73,50	42,89
CFS (10)	Train	57,39	75,94	58,38	85,53	61,11
	Test	50,01	73,57	50,92	85,17	48,05
NB-BACK(44)	Train	88,29	90,42	87,81	91,85	93,82
	Test	39,71	62,23	41,31	75,09	44,22
NB-FOR(1)	Train	84,07	87,82	88,13	91,73	91,23
	Test	51,92	73,57	51,92	85,17	51,92
C4.5-BACK(49)	Train	55,12	75,94	55,12	85,53	55,12
	Test	41,35	63,61	39,76	78,79	45,11
C4.5-FOR(6)	Train	56,25	75,94	58,21	85,53	58,53
	Test	52,24	73,57	53,54	85,17	52,26
SVM-RFE(20)	Train	82,60	78,04	86,01	86,02	86,03
	Test	39,70	71,27	37,09	85,17	40,63
SVM-RFE(30)	Train	85,86	85,48	88,62	91,68	91,23
	Test	36,42	64,10	33,64	80,95	41,33

Table 3. C4.5 Mean Accuracy Results for Waste treatment related Behavior. Number of features selected by each FS method in parenthesis.

FS Method	Acc	Discretization Method				
		No-Disc	5 bins		3 bins	
			EWD	EFD	EWD	EFD
No-FS (68)	Train	86,17	87,46	88,15	94,56	90,76
	Test	22,40	35,64	31,53	40,38	39,21
CFS (8)	Train	37,09	45,97	37,09	49,61	42,01
	Test	34,84	47,18	34,84	49,44	47,09
NB-BACK(27)	Train	84,52	86,97	82,88	91,62	91,11
	Test	27,62	44,51	27,92	42,86	38,56
NB-FOR (2)	Train	44,98	55,18	45,96	59,14	55,83
	Test	40,94	54,23	39,63	55,20	43,47
C4.5-BACK (53)	Train	84,66	87,32	88,64	92,92	91,75
	Test	25,92	36,42	27,65	44,85	37,97
C4.5-FOR (2)	Train	46,80	55,20	46,95	59,80	53,55
	Test	42,23	55,54	42,89	58,13	44,77
SVM-RFE (20)	Train	80,72	83,22	82,85	88,16	87,49
	Test	25,67	33,83	27,58	42,20	38,06
SVM-RFE (30)	Train	84,00	85,32	84,34	90,28	89,47
	Test	25,12	34,73	30,80	41,57	39,35

Table 4. Number of leaves (Lv), features (Fs) and tree size, i.e. number of branches, for the decision tree with the best test accuracy (TsAcc) for each behavior and combination of adequate FS method and EWD-3bins

Method	Mobility				Energy				Waste			
	Lv	Fs	Size	TsAcc	Lv	Fs	Size	TsAcc	Lv	Fs	Size	TsAcc
NO-FS-No-Disc	27	25	53	45,31	23	19	45	40,62	31	23	61	29,03
NO-FS-EWD3	21	17	41	65,62	12	8	23	81,25	23	19	45	49,18
NB-Back-EWD3	5	3	9	**68,75**	11	10	21	78,69	27	19	53	46,77
C4.5-Back-EWD3	20	16	39	62,50	4	3	7	80,65	25	16	49	50,82
SVM-RFE-EWD3	20	10	39	64,06	10	8	19	**83,60**	23	14	45	**51,67**

5 Conclusions

The LOCAW project has investigated sustainable behaviours at the workplace by using, among other techniques, ABM. Decision trees were employed to design the internal behavior of these agents with a dual purpose: to be based on actual data and to be reviewed by experts in the domain. In this paper, discretizacion and feature selection techniques are applied before training the decision trees with the aim of obtaining simpler decision trees, i.e., easier to analyze, but with better generalization capabilities. Experimental results for three different work-related practices denoted that there is always a combination of discretization and feature selection methods that significantly improves the accuracy compared to not using these methods, besides obtaining a much simpler decision tree. As future work, trying to overcome the unbalanced situation, it is proposed to reduce these problems of multiclass classification to multiple binary classification problems using different strategies.

Acknowledgement. This work was supported in part by the European Commission under project "LOw Carbon At Work: Modelling agents and organization to achieve transition to a low carbon Europe", 7th Framework Programme, ENV.2010.4.3.4-1 Grant Agreement 265155 and by the Consellería de Cultura, Educación e Ordenación Universitaria of the Xunta de Galicia through the research grant GRC2014/035, partially funded by FEDER funds of the European Union.

References

1. Ehrentreich, N.: Agent-based modeling. Lecture Notes in Economics and Mathematical Systems 602 (2008)
2. Macal, C.M., North, M.J.: Tutorial on agent-based modelling and simulation. Journal of Simulation 4(3), 151–162 (2010)
3. Sánchez-Maroño, N., Alonso-Betanzos, A., Fontenla-Romero, O., Brinquis-Núñez, C., Polhill, J., Craig, T., Dumitru, A., García-Mira, R.: An agent-based model for simulating environmental behavior in an educational organization. Neural Processing Letters (2015)
4. Macy, M.W., Willer, R.: From factors to actors: Computational sociology and agent-based modeling. Annual Review of Sociology, 143–166 (2002)
5. Quinlan, J.R.: C4. 5: programs for machine learning, vol. 1. Morgan Kaufmann (1993)
6. Hall, M., Frank, E., Holmes, G., Pfahringer, B., Reutemann, P., Witten, I.H.: The weka data mining software: An update. ACM SIGKDD Explorations Newsletter 11(1), 10–18 (2009)
7. Bolón-Canedo, V., Sánchez-Maroño, N., Alonso-Betanzos, A.: A review of feature selection methods on synthetic data. Knowledge and Information Systems 34(3), 483–519 (2013)
8. Hall, M.A.: Correlation-based feature selection for machine learning. PhD thesis, The University of Waikato (1999)
9. Guyon, I., Weston, J., Barnhill, S., Vapnik, V.: Gene selection for cancer classification using support vector machines. Machine Learning 46(1-3), 389–422 (2002)
10. Liu, H., Hussain, F., Tan, C.L., Dash, M.: Discretization: An enabling technique. Data Mining and Knowledge Discovery 6(4), 393–423 (2002)
11. Yang, Y., Webb, G.I.: Proportional k-interval discretization for naive-bayes classifiers. In: Flach, P.A., De Raedt, L. (eds.) ECML 2001. LNCS (LNAI), vol. 2167, pp. 564–575. Springer, Heidelberg (2001)
12. Chawla, N.V., Bowyer, K.W., Hall, L.O., Kegelmeyer, W.P.: Smote: Synthetic minority oversampling technique. Journal of Artificial Intelligence Research 16, 321–357 (2002)

A Simulation of Householders' Recycling Attitudes Based on the Theory of Planned Behavior

Andrea Ceschi[1,*], Ksenia Dorofeeva[1,*], Riccardo Sartori[1], Stephan Dickert[2,3], and Andrea Scalco[1]

[1] University of Verona, Department of Philosophy, Education and Psychology, Verona, Italy
{andrea.ceschi,ksenia.dorofeeva, riccardo.sartori,andrea.scalco}@univr.it
[2] WU Vienna University of Economics and Business
[3] Linköping University, Linköping, Sweden
stephan.dickert@wu.ac.at

Abstract. What would induce householders to recycle their waste in a practicable way? This paper investigates the determinants of recycling behavior through the development of an *Agent-Based Model* (ABM) simulation. Specifically, this contribution examines the effectiveness of a simulation based on different recycling behaviors to better understand the phenomenon. The perspective used takes into consideration the recycling attitude and the sensitivity to social norms based on the *Theory of Planned Behavior* (TPB) to predict recycling outcomes. As a whole, this paper highlights the dominance of dynamics and interacting aspects of waste management for the formulation of effective recycling public policies based on behavioral aspects. In addition, it illustrates how it is possible to use such empirical models as the *Structural Equation Models* (SEM) as a starting point for developing ABM simulations that are closer to both theory and reality.

Keywords: Waste management, Theory of Planned Behavior, Structural Equation Models, Agent-Based Model.

1 Introduction

Environmental protection ranks very high in the global agenda. The waste management sector is expected to achieve significant results in the near future with a significant reduction of the adverse effects of waste on the environment. However, the increasing complexity of the current waste management systems coupled with the demanding environmental protection targets makes the optimization of the waste management strategies and policies challenging. For this reason, waste prevention is the most desirable option, followed by the preparation of waste for reuse, recycling, upcycling and other recovery, with disposal (such as landfill) as the last resort. Low household participation is the key factor complicating the waste recycling scenario in most countries today.

[*] The first two authors contributed equally to the manuscript.

J. Bajo et al. (eds.), *Trends in Prac. Appl. of Agents, Multi-Agent Sys. and Sustainability*, Advances in Intelligent Systems and Computing 372, DOI: 10.1007/978-3-319-19629-9_20

Given these reasons, an arising question is what would induce households to recycle their waste in a practicable way. One of the possible answers lies in a simple psychological phenomenon that is widely known but poorly understood: people's behavior is largely shaped by the behavior of those around them. In psychology, this phenomenon takes the name of *social norms*. Social norms are in fact one of the most powerful customary rules that govern behavior in groups and societies.

Research shows that social norms are most compelling when people are shown evidence that the behavior they are being encouraged to adopt is already practiced by people similar to them (see the *Social Comparison Theory*) (Festinger, 1954). Social norms work primarily at a subconscious level. This means, for example, that people responding to a marketing campaign are mostly oblivious to what drove their behavioral change. Traditional forms of market research, such as focus groups and surveys, are of limited use in a social norm campaign. When people are polled, they typically underestimate the effects of the campaign, because they are not usually aware that it had an effect on them. It has been demonstrated, for example, that the activation of a descriptive social norm encourages prosocial behaviors in citizens (Cialdini, Reno, & Kallgren, 1990). An issue that has received very little attention by literature deals with the question of what is the most effective way to activate policy strategies in order to produce behavioral change. Therefore, to simulate possible scenarios for policy strategies, we created an *Agent-Based Model* ABM) that represents a virtual society.

2 The *Theory of Planned Behavior* and the Recycling Issue

Cognitive psychological modeling can provide the means by which it becomes possible to identify the driving forces behind the recycling behavior and to determine the most likely success factors for public policies. Literature indicates that environmental attitudes and situational and psychological variables are likely to be important predictors of the recycling behavior. Further, the *Theory of Planned Behavior* (TPB; (Ajzen, 1991) provides a theoretical and cognitive framework to understand and explain the influence of these factors.

The TPB was developed from the previous *Theory of Reasoned Action* (Fishbein, Ajzen, 1981) and assumes that people have a rational basis for their behavior in that they consider the implications of their actions. According to the TPB, intentions to engage in recycling behavior are derived from three main factors: 1. subjective norms, 2. attitudes and 3. perceived control.

The concept of *subjective norms* refers to the individual's belief that people important to the decision maker see their behavior as the appropriate way to act. Aceti (2002) argues that people are motivated to recycle by the actual pressure they receive from family and friends to do so. Furthermore, simply knowing that family, friends and neighbors participate in recycling activities increases the likelihood of participation. In this spirit, Stern, Dietz and Kalof (1995) stressed the importance of considering the social structure within which individuals are embedded, based on the belief that social structures shape individuals' experiences and ultimately their personal values, beliefs and behaviors. Following Trafimov and Finlay (1996), it may be suggested that subjective norms are relevant only for participants with higher accessibility of collective self. However, according to Cialdini's *Theory of Normative Behavior* (Cialdini et al., 1990), it may be suggested that the actual impact of subjective social norms is underestimated when it is measured by means of anonymous questionnaires completed in private settings (Stiff & Mongeau, 1994). In fact, Cialdini et al. (1990) showed that, in experimental settings, where an injunctive antilittering norm was made salient, participants' littering behavior was significantly reduced. Following Cialdini and Trost (1998) suggestions, institutions wanting "to activate socially beneficial behavior should use procedures that activate injunctive social norms, since these norms appeared to be more general and more cross-situational effective" (Mannetti, Pierro, & Livi, 2004).

The concept of *attitudes* refers to the individual's evaluation of the action. Boldero (1995) found that intentions to recycle newspapers directly predicted actual recycling and that attitudes toward recycling predicted the recycling intentions. Davies, Foxall, and Pallister (2002) argue that recycling attitudes should be separated into two components: 1. *affective*, representing feelings about recycling; 2. *cognitive*, representing knowledge of outcomes and consequences of performing behaviors (Tonglet, Phillips, & Read, 2004).

The concepts of *perceived control and moral obligation* refer to the individual's perception of their ability to perform behaviors. Taylor and Todd (1995) found that both attitudes toward recycling and perceived behavioral control were positively related to individuals' recycling and composting intentions. According to the TBP, perceived behavior control will influence actual behavior only if the behavior is not completely under the person's volitional control.

3 A Method for Modelling: Connecting Theories and Models

Agent simulations range from highly structured artificial worlds with few simple rules and constraints (Kohler & Gummerman, 2001) to more complex models in which agent interactions constrain subsequent iterations of the simulation (Sawyer, 2001) and/or multiple structural layers are considered (Stinchcombe, 2001). It is well known that the development of these algorithms is the most fragile aspect of the simulation analysis. In the specific case of this paper, in order to design social networks the crucial preference is the identification of an amount of the agent's attributes that are significant for recycling behavior. They span from basic demographic attributes (i.e., age, education and income), to more specific features (i.e., environmental sensitivity, self-confidence and sense of social belonging) (Ceschi, Rubaltelli, & Sartori, 2014).

These attributes have the greatest impact and therefore it is important to consider them for the aims of the analysis. For this reason, it is recommended to start from some empirical models, such as *Structural Equation Models* (SEM), which are widespread in social and psychological science and used to represent human behaviors. They consist of equations that link latent variables and their indicators by using exogenous causes. A remarkable benefit of this framework is that correlations of observed indicators are clearly made as arising out of subjacent factors that are accountable for the results (Krishnakumar & Ballon, 2008). That is, SEMs are able to reveal and to quantify the relationship between a behavioral expression and its underlying psychological construct. Apart from this, results from SEM are usually not linked to individual differences. This is a limit of this quantitative method. Essentially, individual differences are characterized as a set that makes individuals particular, according to their inclinations, capabilities and outcomes. Starting from here, ABM can help to dynamically represent, in a natural way, several scales of analysis and the importance of structures at different levels, none of which is easy to accomplish with other modeling techniques (Gilbert & Terna, 2000).

In the current paper, the aim of the analysis is to present a model able to simulate a number of characteristics that have been scaled from the original work of Chu and Chiu (2003), modeled and assigned with probability distributions to simulate the recycling behavior. Usually, the purpose of this stochastic effort is to endow agents with a 'personality'. Contemplating the possibility of fuzzy logic implies greater simulation realism as different agents act differently in the same situation. Agents with personality lead to the modeling of more complex interactions where, for example, hypotheses may be tested more effectively by considering teams of agents with different personalities rather than single agents (Garson, 2009).

4 The Planned Recycling Agent Behavior Model

Our analysis is based on a simulation model of the *Planned Recycling agent Behavior (PRB_1.0)*[1] that produces virtual neighborhoods with different agent types, waste generation and collection processes (Fig. 1). The scaling of the agents' features is based on coefficients relating to the TPB and taken from a SEM on motivations to recycling behavior developed by Chu and Chiu (2003). The application of scaling allows us to accelerate the simulations lowering hardware requirements to run the simulation, leaving untouched the original ratio between agents' variables. The SEM shows four crucial coefficients defining the recycling behavior: the environment attitudes (ea), the subjective social norms (ssn), the perceived behavioral control and the moral obligation constructs. In the model, the last two factors compose a one-of-a-kind factor: the b&m. The SEM values of the constructs described are set by Bayesian computation, and utilized in the simulation as probabilistic variables. To recreate the city, the algorithm of the simulation consists of an open source code and it is based on three types of agents (see Fig. 1):

The *neighborhood agent* – All neighborhood agents generate recycled rubbish (Rre) and not recycled trash (R). This is based on the probabilities of psychological constructs and other agent habits. Neighborhood agents recycle if they possess high levels of environment attitudes (ea) and high subjective social norms (ssn). This link is not mediated by other aspects (Fig. 2). Probabilities of these psychological constructs are normally allocated among agents. A neighborhood agent can furthermore recycle if behaviorally influenced (BI) by other agents or by the environment.

As for ea and ssn, the coefficient of being behavioral affected is extracted from Chu and Chiu (2003) model and it consists of a unique coefficient composed of the perceived behavioral control and the moral obligation (b&m) constructs.

---

[1] The full model, code and documentation is available via the www.openabm.org website.

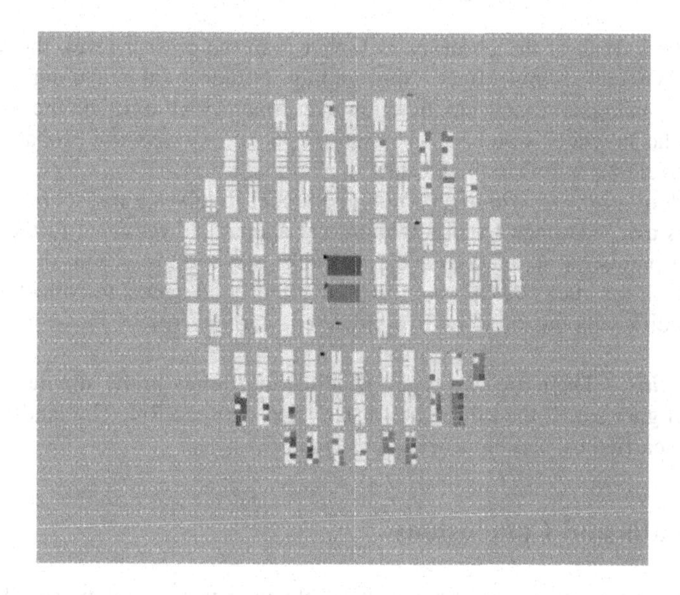

Fig. 1. Example of the PRB_1.0 simulation. The simulation presents three different types of agents: 1. Neighborhood agents: they turn their color from yellow to red color when R level reach the critical level of R, the orange color of agents represents a state of agents close to the critical level, 2. Garbage transportation system represented as rectangles in the model, 3. Landfills: the green rectangle and grey rectangle represent the recycling and non–recycling landfills.

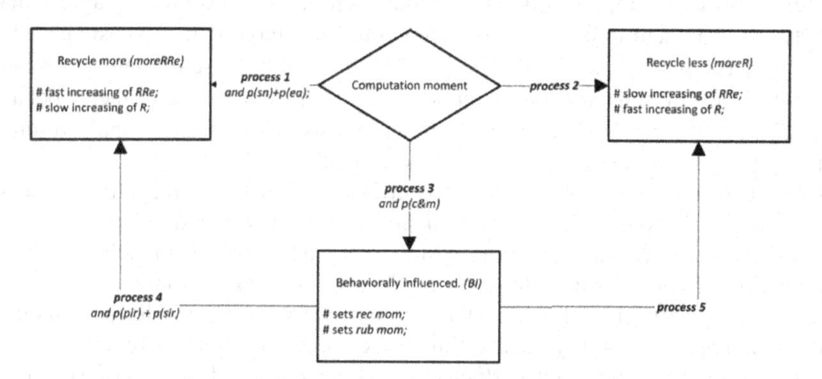

Fig. 2. Schema of reasoning of PRB_1.0 agents

When an agent is BI, it randomly detects a neighbor agent close to its radius and is more available to generate Rre than R if the neighbor watched is recycling. On the contrary, an agent will be more disposed to generate R if neighbors are not recycling.

We define "peer influence" in recycling (pir) the predisposition of an agent to be motivated by others. The inclination to recycle over time is calculated by a decay (and an inverse decay) function formulated starting from the conceptual model of motivation and satisfaction of needs over time (Jager & Janssen, 2012) based on the characteristics of human sensitivity. Additionally, the environment influences the recycling behavior (through pir). Through the observation, agents also examine the R level of others. If R of the observed is higher than the critical level, the agent ends recycling. We called this effect "surrounding influence in recycling" (sir). It is calculated by using another decay function over time, which relies on the volume of R present in the neighbor watched. The sum of pir and sir sets the probability for an agent to recycle when it is in the BI state.

The *garbage transportation system* – The model involves a transportation system, which takes away garbage from neighborhood agents and moves it to the collecting points. The pathways adopted by pick-up trucks are optimized considering distance and time. Pick-up trucks get to the closest neighborhood agents to collect R and Rre. After a specific amount of R or Rre collected, garbage trucks move to the closest landfill.

The *landfills* – There are two types of collecting points in the simulation: one for unseparated garbage R, the other one for recycled garbage Rre. The landfill removes the garbage carried by pick-up trucks over time.

5 Results and Conclusions

Preliminary results of the model are available on *www.openabm.com.* They show stability and reliability in relation to the outcomes of the simulation. The visual impact created a virtuous circle where household motivation to recycle is reinforced. This circle expresses the consequences of descriptive social norms. On the contrary, the failure in recycling when the environment is full of rubbish contaminates the neighbors' behavior (Fig. 3). Literature about social norms and littering agrees that in a 'dirty' environment individuals are inclined to litter more than those subjected to a 'clean' environment (e.g., Cialdini et al., 1990), mainly because of the peer influence, due to the fact that agents continuously observe and mimic each other's behavior. Similarly, the surrounding has its own effect because the amount of garbage present in the system drives the trend away from its stable level.

Agent-based models can simulate the efficacy of different recycling campaigns under equal conditions and, at a subsequently stage, allow the simulation of specific policies under different conditions. Moreover, agent-based models are mostly structured on algorithms that illustrate the behaviors of agents, identify their causal effects and specify critical parameter estimates. Therefore, stochastic simulation, while retaining its versatility, is also time-effective and cost-effective. It is important to state that the agent behavior is stochastic. As we suggested, factors of SEM can be implemented based on ABMs, in contrast to equations of aggregation. Furthermore, the ABM approach for modeling is versatile, in particular in relation to spatial simulations in mobility models that are agent-based. An agent-based model is usually characterized within a given organization or environment. Through this simulation, researchers can identify causal effects, specify critical parameters, and clarify how recycling processes evolve over time, in order to propose policies that are more effective.

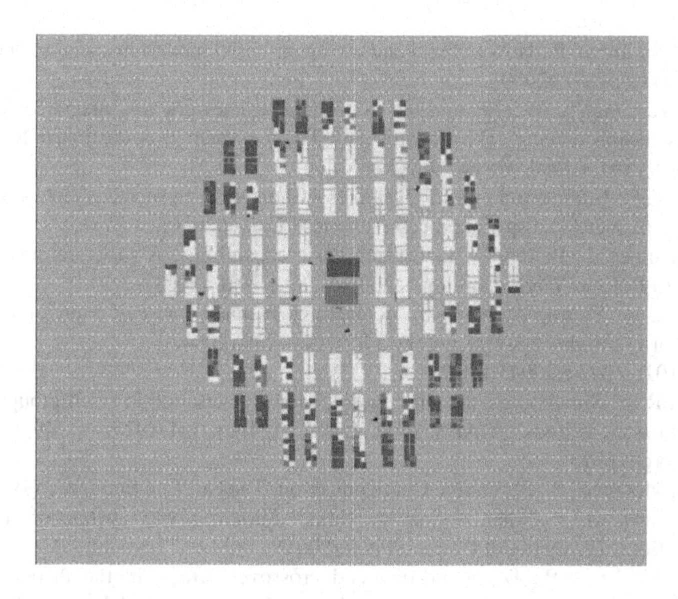

Fig. 3. A figure from the ABM based on the PRB_1.0 with higher R levels than Fig. 1. Neighborhood agents turn color because of the R level. When R is equal to the critical level they turn red, orange if they are close to the critical level, yellow is the R situation is stable.

References

1. Aceti, J.: Recycling: Why People Participate. Why They Don't, Massachusetts Department of Environmental Protection, Boston, MA (2002)
2. Fishbein, M., Ajzen, I.: Attitudes and voting behavior: An application of the theory of reasoned action. Progress in Applied Social Psychology 1, 253–313 (1981)
3. Ajzen, I.: The theory of planned behavior. Organizational Behavior and Human Decision Processes 50(2), 179–211 (1991)
4. Ceschi, A., Rubaltelli, E., Sartori, R.: Designing a Homo Psychologicus More Psychologicus: Empirical Results on Value Perception in Support to a New Theoretical Organizational-Economic Agent Based Model. In: Omatu, S., Bersini, H., Corchado Rodríguez, J.M., González, S.R., Pawlewski, P., Bucciarelli, E. (eds.) Distributed Computing and Artificial Intelligence 11th International Conference. AISC, vol. 290, pp. 71–78. Springer, Heidelberg (2014)
5. Chu, P.Y., Chiu, J.F.: Factors Influencing Household Waste Recycling Behavior: Test of an integrated Model1. Journal of Applied Social Psychology 33(3), 604–626 (2003)
6. Cialdini, R.B., Reno, R.R., Kallgren, C.A.: A focus theory of normative conduct: recycling the concept of norms to reduce littering in public places. Journal of Personality and Social Psychology 58(6), 1015 (1990)
7. Davies, J., Foxall, G.R., Pallister, J.: Beyond the intention–behaviour mythology an integrated model of recycling. Marketing Theory 2(1), 29–113 (2002)
8. Festinger, L.: A theory of social comparison processes. Human Relations 7(2), 117–140 (1954)
9. Garson, G.D.: Computerized Simulation in the Social Sciences A Survey and Evaluation. Simulation & Gaming 40(2), 267–279 (2009)

10. Gilbert, N., Terna, P.: How to build and use agent-based models in social science. Mind & Society 1(1), 57–72 (2000)
11. Jager, W., Janssen, M.: An updated conceptual framework for integrated modeling of human decision making: The Consumat II. Paper presented at the Paper for Workshop Complexity in the Real World@ ECCS (2012)
12. Kohler, T.A., Gummerman, G.J.: Dynamics of human and primate societies: agent-based modeling of social and spatial processes. Oxford University Press (2001)
13. Krishnakumar, J., Ballon, P.: Estimating basic capabilities: A structural equation model applied to Bolivia. World Development 36(6), 992–1010 (2008)
14. Mannetti, L., Pierro, A., Livi, S.: Recycling: Planned and self-expressive behaviour. Journal of Environmental Psychology 24(2), 227–236 (2004), doi:10.1016/j.jenvp.2004.01.002
15. Sawyer, R.K.: Simulating emergence and downward causation in small groups. In: Moss, S., Davidsson, P. (eds.) MABS 2000. LNCS (LNAI), vol. 1979, pp. 49–67. Springer, Heidelberg (2001)
16. Stiff, J., Mongeau, P.: Persuasive Communication. The Guilford Press, NY (1994)
17. Stinchcombe, A.L.: When formality works: Authority and abstraction in law and organizations. University of Chicago Press (2001)
18. Taylor, S., Todd, P.: Decomposition and crossover effects in the theory of planned behavior: A study of consumer adoption intentions. International Journal of Research in Marketing 12(2), 137–155 (1995)
19. Tonglet, M., Phillips, P.S., Read, A.D.: Using the Theory of Planned Behaviour to investigate the determinants of recycling behaviour: a case study from Brixworth, UK. Resources, Conservation and Recycling 41(3), 191–214 (2004)

Introducing LCA Results to ABM for Assessing the Influence of Sustainable Behaviours

Tomás Navarrete Gutiérrez, Sameer Rege, Antonino Marvuglia, and Enrico Benetto

Luxembourg Institute of Science and Technology (LIST)
41 rue du Brill, 4422 Belvaux, Luxembourg
{tomas.navarrete,sameer.rege,
antonino.marvuglia,enrico.benetto}@list.lu

Abstract. The paper is focused on the application of ABM to simulate the evolution of the agricultural system of the Grand Duchy of Luxembourg under the external shock of fostering the production of maize to be used for bio-methane generation. The modeling exercise is part of a larger scale effort, which aims at the evaluation of the potential environmental impacts arising from policy implementation, following the methodology known as Consequential Life Cycle Assessment (CLCA). The application of ABM brings from one side an additional level of complexity to the assessment, but it also allows a finer scale bottom-up modeling, allowing the simulation of aspects (such as behavioral components in farmers' choices or adaptive social processes such as imitation) that was not possible to grasp from other (e.g. purely economically oriented) modeling perspectives. The ABM model implemented has as many agents as the inventoried farmers in Luxembourg are. Although the paper only presents preliminary results, it already allows exploring the influence of farmers' environmental awareness on the environmental impacts linked to farming activities. This is possible thanks to the attribution to the agents' profiles of one specific feature which simulates their "green consciousness level".

1 Introduction

Life cycle assessment (LCA) is a standardized methodology used to quantify the environmental impacts of products across their whole life cycle [8] which is nowadays recognized worldwide, although with some persisting limitations and certain barriers, as witnessed by the survey of [3]. It has been used at the fine grain level (at the product level) or at a more global level (policy).

In this latter case (policy evaluation) a more pertinent methodology that has nowadays gained recognition is the so-called Consequential LCA (CLCA). CLCA aims at evaluating the direct and indirect environmental consequences of a strategic decision taken in a moment t0, which can be in the past, present or future. For example, in the specific case of agro-systems, the land use changes and related consequences following a bio-energy policy. The ultimate aim of

J. Bajo et al. (eds.), *Trends in Prac. Appl. of Agents, Multi-Agent Sys. and Sustainability*,
Advances in Intelligent Systems and Computing 372, DOI: 10.1007/978-3-319-19629-9_21

the study is then evaluating the difference (delta) in terms of environmental inventory (and related impacts) between the two time marks set for the simulation (t0 and t0+n). In the specific case of agro-systems, economic modeling approaches, assuming supply-demand equilibrium, are increasingly applied to derive the linkages between agricultural operations and economic market, and their consequences in terms of environmental damage costs and social costs. However, behavioral criteria and adaptive social processes (e.g. imitation) are not taken into account in top down economic models and basing the modeling exercise exclusively on economic criteria is likely to turn out inappropriate to forecast structural changes, simply because economic decisions are never purely rational.

The MUSA (MUlti agent Simulation for consequential LCA of Agro-systems) project `http://musa.tudor.lu`, aims to simulate the future possible evolution of the Luxembourgish farming system, accounting for more factors than just the economy oriented drivers in farmers decision making processes. We are interested in the behavioral aspects of the agricultural systems, including the green consciousness of the farmers. The challenges facing the farming system are manifold. Dairy and meat production have been the financial mainstay of the national agricultural landscape with production of cereals and other crops as a support to the husbandry. This sector is fraught with multiple rules and regulations including restrictions on milk production via quotas as dictated by the Common Agricultural Policy of the EU. There is also a complex set of subsidies in place to enable the farmers to be more competitive. The quotas are set to disappear adding to increased pressure on the bottom line for dairy farmers. Just as a chain is as strong as its weakest link any model is as robust as its weakest assumption. Model building is a complex exercise but modelling behaviour is far more complex. Statistical and optimization models fail to account for vagaries of human behaviour and have limited granularity. Preceeding the MUSA project, in [11] we have built a partial equilibrium model for Luxembourg to conduct a consequential LCA of maize production for energy purposes (dealing with an estimated additional production of 80,000 t of maize) using non-linear programming (NLP) and positive mathematical programming (PMP) approaches. PMP methodology [7] has been the mainstay of modelling methodology for agriculture models relating to cropping patterns based on economic fundamentals. This approach converts a traditional linear programming (LP) into a NLP problem by formulating the objective function as a non-linear cost function to be minimised. The objective function parameters are calibrated to replicate the base case crop outputs. This approach is useful as a macro level such as countries or regions where one observes the entire gamut of crops planted at a regional level. To increase the granularity to investigate the impacts of policy on size of farms is still possible provided each class of farms based on size exhibits plantation of all crops in the system. When the granularity increases to the farm level, the crop rotation takes priority for the farmer and then one observes only a subset of the entire list of crops in the system. It is at this level that the PMP approach fails as the objective function is calibrated to the crops observed at a specific point in time. This limitation is overcome by the agent based model (ABM) approach. To investigate the possible set of outcomes due to human responses to financial and natural challenges, ABM are a formal mechanism to validate the range of outcomes due to behavioural differences.

LCA can produce as outputs different levels of environmental assesments. The levels are termed mid-point and end-point in the LCA jargon. They correspond respectively to assessments given at a semi-aggregated and aggregated level. An example result from an LCA assessment is the now famous "Carbon Footprint".

We consider that LCA assessments is a valuable tool to inform farmers on the potential (in the LCA terms) impact of their activities. Starting from this, we focus on the question of modelling the inclusion of LCA results in the behaviour of farmers in ABM model.

In the work we report here, we investigate how to introduce the results of LCA into agent-based models to simulate the evolution of an agricultural system. The main question remains how to use the specific results of an initial LCA within the decision process of farmers. For our work, we shall use our own initial agent-based model that will allow agents to perceive the results of an LCA (for the full agricultural system) and use them in the farm planning.

In section 2 we present a brief tour of different agent-based models dealing with agricultural systems as well as LCA research done using agent-based models. Then, in section 3 we describe our model as well as our proposition to link it with LCA. We present the initial results of the simulations in section 5 and present our conclusions in section 6.

2 Literature Survey

[5] use an ABM called AgriPoliS, for simulation of agricultural policies. The focus is on simulating the behaviour of farms by combining traditional optimization models in farm economics with agent based approach for the region of Hohenlohe in southwest Germany. The paper classifies farms into size classes and assumes all farms in a specific class size to have the same area. The variation in the farms is on account of individual asset differences (both financial and physical). Farms are classified only as grasslands or arable lands with no further classification by crops or rotation schemes. [1] is another application on the lines of [5] with an integration of farm based linear programming models under a cellular automata framework, applied to Chile. The model explicitly covers the spatial dimension and its links to hydrology with a policy issue of investigating the potential benefits of joining the Mercosur agreement. [10] is an application of agent based modelling framework to assess the socio-ecological impacts of land-use policy in the Hong Ha watershed in Vietnam. The model uses in addition to farms as agents, the landscape agents that encapsulate the land characteristics. Additional sub models deal with farmland choice, forest choice, agricultural yield dynamics, forest yield dynamics, agent categorizer that classifies the households into a group and natural transition to enable transformation between vegetation types. [2] is a similar study to simulate structural and land use changes

in the Argentinian pampas. The model is driven by agent behaviour that aims to match the aspiration level of the farmer to the wealth level. The farmer will change behaviour until the two are close. In addition to 6 equal sized plots in the farm, the farmer has a choice of planting 2 varieties of wheat, maize and soyabean. The model studies the potential penetration of a particular crop and find that soyabean is being cultivated on a much larger scale.

Despite most of the previously cited related works include a high number of economical aspects in their models, they lack the inclusion of aspects related to green consumption and production.

3 Data and Model Structure

In the model, the farmers are represented via entities called "agents" who take decisions based on their individual profiles and on a set of decision (behavioural) rules defined by the researchers on the basis of the observation of real world (e.g. the results of the questionnaire) and on the interactions with other entities. The model currently includes one reactive rule regarding the change of crops planted at the farm.

Luxembourg provides statistics about the economy of the agricultural sector through STATEC [13] and Service d'Economie Rural [12]. The statistics deal with the area under crops over time, farm sizes and types of farms by output classification, use of fertilizers, number of animals by type (bovines, pigs, poultry, horses) and use (meat, milk). The latest year for which one obtains a consistent data set across all the model variables was 2009[1]. In 2009 there were 2242 farms with an area of 130,762 ha under cultivation that included vineyards and fruit trees and pastures amongst other cereal and leaf crops.

Table 1 shows the distribution of farm area by size of farms in Luxembourg for the year 2009. We also know the total area of crops planted under each farm type. Figure 1 shows the initial proportions of each crop present in 2009. From [9] we know the different rotation schemes for crops. Rotation schemes are used by farmers to maintain the health of the soil and rotate cereals and leaves on the same field in a specified manner. We randomly assign a rotation scheme to each farm and then randomly choose a crop from cereals or leaves. As to crops that are neither cereals nor leaves, they are a permanent cultivation such as vineyards, fruits, meadows or pastures. Once the random allocation of crops has been made to the farms, we scale the areas so as to match the total area for the crop under a specific farm type.

3.1 Model Structure

The model is intended to be used as part of a LCA. The LCA measures the (environmental) consequences (at a global scale and in a life cycle perspective) of different decisions made by farmers in Luxemburg. The model answers to the

[1] The details of the data are available in [11].

Table 1. Distribution of Farms by Size (ha) in Luxembourg in 2009

	Farm Class	A	B	C	D	E	F	G	H	I
	TOTAL	< 2	2-4.9	5-9.9	10-19.9	20-29.9	30-49.9	50-60.9	70-99.9	100+
Number	2242	230	165	217	186	116	246	263	398	421
Area (ha)	130762	131	598	1533	2667	2890	9956	15743	33583	63661
Average size	58.32	0.57	3.62	7.06	14.34	24.91	40.47	59.86	84.38	151.21

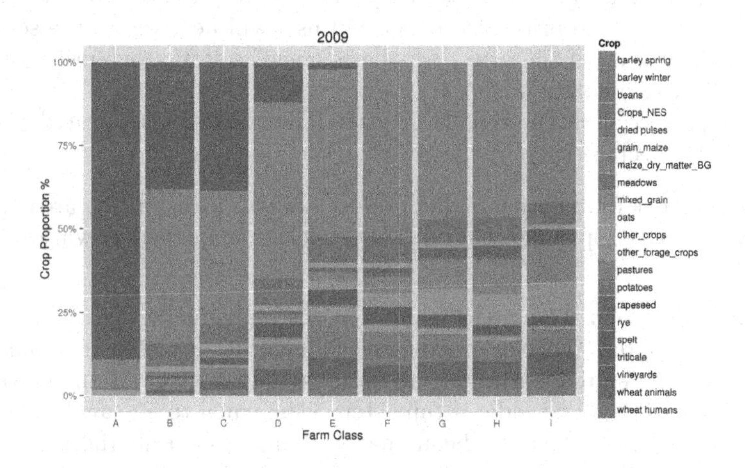

Fig. 1. Initial proportions of crops planted in 2009

question: "What are the (number of) products (crops and animals) being produced in Luxembourgish farms, at a given time?". More specifically, we would like to answer this question under the condition that a certain policy-driven scenario is put in place. In order to do so, the model provides as output the changes (in terms of hectares of land planted with each crop) in the land use occurring as a consequence of the exogenously imposed change (for example the decision to reach an additional production of 80,000 t of maize). In a future enhanced version of the model, it will also be able to provide the changes in terms of cattle heads, milk and meat production. Once these changes are computed, they are used as inputs for the LCA calculations. The answers to this question can then be used to perform an environmental assessment taking into consideration the changes over time of the products of each farm.

The ABM represents the agricultural system of Luxembourg, it is built around the activity of farmers in the country and includes the following entities.

Farmers. The autonomous agents of the system.

Farms. This is a container object used to organize the model. Farms in the model are instantiated using the above mentioned statistics.
The current model of the behaviour of the farmers is to consider that they follow a rational expected maximization behaviour. This means that a farmer will decide to change the crops to plant, based on the yield of a given crop, the price for the last season of the crop and the costs of the crop. The farmer

agent, will decide to change a crop in his farm, that suits the rotation scheme, taking the list of available crops from the environment.
Each farm has an associated rotation scheme. Along with initial allocation of a rotation scheme per farm, we set up an initial allocation of crops planted in a given proportion in the farm.

Product Buyers. In the model, a buyer will offer to buy the production for the farmers. There is only one buyer that will offer to buy every product of the whole country. The prices the buyer will use, will be given in the scenario. The specification of the prices for the moment is static: one price for each "year" in the simulation.

Crop. Crops can be either cereals or leafs. They have an associated yield in tonnes per hectare.

For the current formulation of the project, we will focus on the area (in ha) planted with each Crop, for each farm. The model accounts the following different scales.

Description Levels. The model is given at the individual level, for farms and farming resources. For the prices, we will consider a global representation where all prices are identical for all farmers and are given by the market.

Time. The model will consider a time step being equal to a year. In one time step, the agent sows at the beginning, then harvests, sells the production, and finally decides to change or not a specific crop for this rotation scheme (substitute a **C**ereal or a **L**eaf).

Space. Each farm has a size (in ha) of arable land. No GIS information is included in the model for the moment.

Designing Concepts. Agents follow a reactive architecture. They observe the prices in the system and change parameters of their main behaviour accordingly. Since there are no interactions between agents, we consider a sequential application of behaviours.

Implementation. In order to keep a maximum flexibility during the development of the model as well as to be able to use the outputs of the model as the inputs for the LCA, we have built our own simulator from scratch using java as the programming language.

3.2 Connection to LCA

The interested reader may look at [11] for the details of the LCA model created for the environmental assessment of the luxembourgish agricultural system.

3.3 Model Interconnection

In LCA, studying a particular functional unit requires some modelling.

A block diagram made with SysML of the previous description is given in figure 2. Any LCA software can be used to perform the environmental life cycle impact assessment (LCIA) phase.

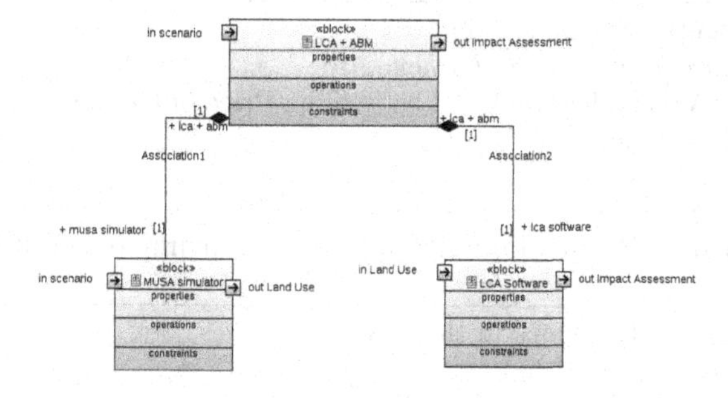

Fig. 2. Block description diagram of the system including the ABM simulator and the LCA software. The block called "ABM+LCA" is in charge of distributing the inputs to the ABM simulator (block "MUSA simulator") and pass the outputs (Land Use) to the LCA software to obtain the final results (Impact Assessment).

4 Simulation Workflow

4.1 Global Simulation

ABM models are simulation models, and as such they must be simulated before they can actually answer the questions they were built for. The simulator that implements the ABM has the following workflow.

1. Initialize the farm sizes.
2. Assign to each farm a random rotation scheme.
3. Assign to each farm a set of crops being planted (depending on the type of rotation scheme).
4. Scale the sizes of the farms and the fields being planted with each crop to match the statistics of 2009.
5. Apply the behaviours associated to each agent in a "pre-market" phase.
6. Let the agents sell their produce (the result of the yield per hectare per crop) in a market with prices estimated form historical records and future forecasts made using holt winters time series technique.
7. Apply the behaviours associate to each agent in a "post-market" phase.

4.2 Behaviours

We currently have implemented two different behaviours: one that represents a use of exclusively economic drivers in the decisions (deemed of profit maximazation) and another one where the most important criterion for taking decisions is the environmental impact of planting a given crop (in terms of CO_2 emissions).

The focus on CO_2 is presented here as a mere example, given the fact that the environmental impacts we calculate are the entire set of so called midpoint impacts obtainable with the life cycle impact assessment (LCIA) methods commonly used in LCA. In particular, we applied the consensus method ReCIPe [1]. In order to calculate the potential environmental impacts, the lifecycle amount

```
Estimate potential economic gain;
for Each crop i in the potential scheme do
    gain[i] ← AreaUnderCrop[i] * Last Selling Price of Crop[i];
    total gain += gain[i];
end
```

Algorithm 1. Profit Maximization behaviour implementation.

```
l ← perceive available Crops in the Environment;
for Each Crop c in l do
    c.profit ← calculate potential value (c);
end
for Each "Cereal" crop i currently planted in the farm do
    s ← select highest value(l,currentplantation, i);
    if ∃s then
        substitute i with s;
    end
end
```

Algorithm 2. Generic behaviour for substituting crops in the rotation scheme. For the profit maximization behaviour, the highest value function uses CO_2 emissions as the criteria (the lower the emissions, the higher the value) and for the profit maximization, the potential economic gain is the criteria used.

of each pollutant emitted has to be converted (using conversion factors called "characterization factors") in an equivalent amount of a reference substance, acting on the same impact category (for example CO_2 and CH_4 act on the category "climate change" and CO_2 is the reference substance). Midpoint characterization factors are therefore based on equivalency principles, i.e. midpoint characterization scores are expressed in kg-equivalents (kgeq) of a substance compared to a reference substance. In other words, the kgeq of a reference substance expresses the amount of a that reference substance that equals the impact of the considered pollutant within the midpoint category studied. LCIA results of midpoint categories can be represented in two ways: 1. in *absolute* units (kgeq-substance) or 2. in *relative* units (percentage of impact compared to one reference scenario; for more details see [6].

A third behaviour deemed "green consciousness" is introduced to add heterogeneity in the behaviours. An agent (farmer) has an internal green consciousness (a value between 0 and 1) that dictates the strength of the environmental consdierations for a farmer. If the value is lower than 0.5 the farmer will actually use the same criteria as behaviour "profit maximization" to change the crops for next year. On the other case (green consciousness higher or to 0.5) the criteria to use to select a substitution crop will be the environmental impact of the list of available crops. Algorithms 1 and 2 specify the behaviours.

4.3 Scheduling

At each phase of the simulation (as described in 4.1) the agents are scheduled to be active in a random and synchronous order: one agent at the time is randomly selected and the corresponding behaviours (pre-market, market and post market) are applied to each agent.

5 Experiments

To obtain a wide-ranging perspective of the environmental impacts and damages linked to the described production system, and to check the consistency of the results, several different Life Cycle Impact Assessment (LCIA) methods were applied. The analysis of midpoint impacts, as well as endpoint damages was calculated using the ReCIPe [6] methodology. Just for exemplification purposes, we run 5 simulations with the ABM model and we computed the corresponding variations in the cultivated area of each crop. These variations (deltas) were fed to a LCA software, which allowed the calculation of the corresponding midpoint LCIA results. In Figure 3 we compare the results obtained for the five simulations in absolute units.

We have initial results obtainned by simulating the model using the following parameters.

- Each farmer has a random green consciousness that is taken form a uniform distribution between 0 and 1.
- Each simulation is executed only over one time step (each time step reresenting a year)
- All agents perceive in perfect conditions the available crops in the environment as well as the LCA results.
- Each agent has only one post-market behaviour: green consciousness.

5.1 Results

Figure 3 shows the percentual changes for each of the different categories obtained from the LCA assessment based on the changes between the initial proportions of crops planted in the model, and the resulting after applying the behaviours to each agent. For each impact category, the simulation for which the impacts are the highest is assigned the 100% value and the impacts arising from the other simulations are scaled accordingly.

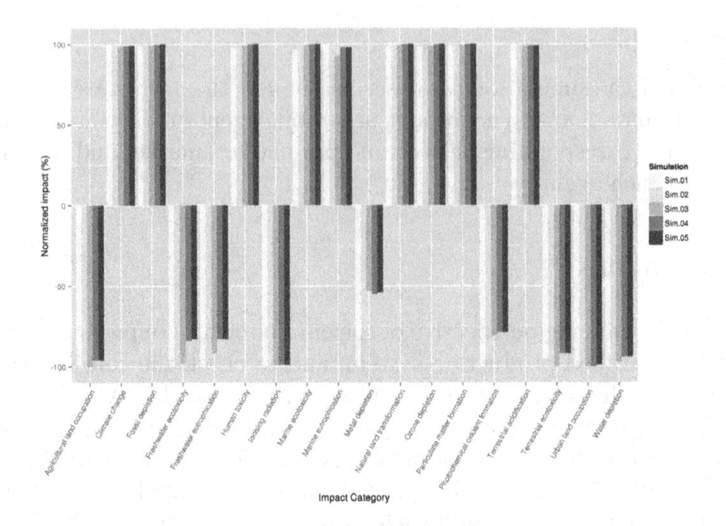

Fig. 3. Relative changes for each environmental indicator for each of the different simulations. Sim 1 through Sim 5 refer to each of the five simulation runs.

6 Conclusions

This paper presents the first version of an ABM which aims at simulating a plausible evolution of the Luxembourgish agriculture and farming system under different conditions (status quo and externally imposed scenarios) at the aim of evaluating the potential environmental consequences of policy driven actions.

We have managed to implement an ABM that lets the agents take decisions based on information they perceive from the environment that is available to all of them. The information was the CO_2 footprint of different crops cultivated in 2009 in Luxembourg. This is an initial solution to the question of how to include the LCA results in an ABM model.

Our initial results, concerning the LCA suggest that if a farmer solely focuses on CO_2 as a criterion and has a "green consciousness" that compels him to make changes in his rotation scheme, he may end up improving his carbon footprint, but may worsen other environmental impacts that could be of higher importance for his specific activities (e.g. those related to soil and water).

For the moment, we consider only the crop type of resources that a farm may have, but from previous experience [14] and from continous discussions with different stakeholders of the Luxembourgish agricultural sector, we know that for highly integrated farming systems, animal breeding is also a very important factor in the decisions at farm planning level. Therefore efforts are ongoing to include this element in the model as well.

6.1 Perspectives

In this first version of the model, we are not taking into consideration the subtleties of the technical orientation of the farms (deemed OTE classification in the european context [4]). Including this would allow us to give a more generalized (at least at the european scale) point of view of our results.

Although we implemented a solution that includes a so-called green consciousness, on how to let the agents react using information from LCA results, the question for future work is how to characterize and let evolve this consciousness. For the matter of the agents profile definition, we are currently conducting a survey to better characterize the decision making processes in farm planning. As for the evolution, we shall include in further refinements of the model, the use of social networks to whom the farmers would belong and let the interactions with members of the networks influence the evolution of the green consciousness. We can at least think of the professional networks (most farmers are professionally adviced by local associations) and the proximity networks (geographical neighborhoods). So far the model fixes the green consciousness of the farmers at the initialization phase, but it is to be expected that this consciousness evolves over time. Further refinements could include social contagion and imitiation as mechanisms to let green consciousness evolve as agents interact with other agents.

Our inclusion of the LCIA results for the moment is limited to executing simulations over one time step and then calculating the LCA results form the outputs of the simulation. In future work, we envision using the results from the LCIA of each time step, and re-inject them in the environment, so that agents can perceive them and also let the green consciousness and decisions evolve based on the evolution of the LCA results. We can think of this as a feedback loop. However, for this we need to develop LCIA tools that can directly be called from the simulator. Our current tools, like for most LCA practitioners, have a limited functionalities that support interaction with other software.

Acknowledgments. Luxembourg's National Research Fund (FNR) is acknowledged for the financial support of project MUSA with id: C12/SR/4011535.

References

1. Berger, T.: Agent-based spatial models applied to agriculture: a simulation tool for technology diffusion, resource use changes and policy analysis. Agricultural Economics 25, 245–260 (2001)
2. Berta, F.E., Podestá, G.P., Rovere, S.L., Menéndez, A.N., North, M., Tatarad, E., Laciana, C.E., Weber, E., Toranzo, F.R.: An agent based model to simulate structural and land use changes in agricultural systems of the argentine pampas. Ecological Modelling 222, 3486–3499 (2011)
3. Cooper, J.S., Fava, J.A.: Life-Cycle Assessment Practitioner Survey: Summary of Results. Journal of Industrial Ecology 10(4), 12–14 (2006)
4. European Commission: COMMISION REGULATION (EC) No 1242/2008 of 8 December 2008 establishing a Community typology for agricultural holdings. Official Journal of the European Union (December 2008)
5. Happe, K., Kellermann, K., Balmann, A.: Agent-based analysis of agricultural policies: an illustration of the agricultural policy simulator AgriPoliS, its adaptation, and behavior. Ecology and Society 11(1), 329–342 (2006), `http://www.ecologyandsociety.org/vol11/iss1/art49/`

6. Heijungs, R., Goedkoop, M., Struijs, J., Effting, S., Sevenster, M., Huppes, G.: Towards a life cycle impact assessment method which comprises category indicators at the midpoint and the endpoint level. Report of the first project phase: Design of the new method VROM report (2003), `http://www.leidenuniv.nl/cml/ssp/publications/recipe_phase1.pdf`
7. Howitt, R.: Positive Mathematical Programming. Amer. J. Agr. Econ. 77, 329–342 (1995)
8. ISO: Environmental management — Life cycle assessment — Principles and framework. ISO 14040:2006, International Organization for Standardization, Geneva, Switzerland (2010)
9. KTBL: Faustzahlen für die Landwirtschaft (in German). Iso, Kuratorium für Technik und Bauwesen in der Landwirtschaft, Darmstadt, Germany (2006)
10. Le, Q.B., Park, S.J., Vlek, P.L., Cremers, A.B.: Land-Use Dynamic Simulator (LUDAS): A multi-agent system model for simulating spatio-temporal dynamics of coupled human–landscape system. I. Structure and theoretical specification. Ecological Informatics 5, 203–221 (2010)
11. Rege, S., Arenz, M., Marvuglia, A., Vázquez-Rowe, I., Benetto, E., Igos, E., Koster, D.: Quantification of agricultural land use changes in consequential Life Cycle Assessment using mathematical programming models following a partial equilibrium approach. Journal of Environmental Informatics (forthcoming)
12. SER: `http://www.ser.public.lu/`
13. STATEC: `http://www.statistiques.public.lu/en/actors/statec/index.htm`
14. Vázquez-Rowe, I., Marvuglia, A., Rege, S., Benetto, E.: Applying consequential LCA to support energy policy: Land use change effects of bioenergy production. Science of the Total Environment 472, 78–89 (2014)

Part VI

Special Session on Emotional Software Agents (SSESA) + Intelligent Educational Systems (SSIES)

Exploring Selfish *versus* Altruistic Behaviors in the Ultimatum Game with an Agent-Based Model

Andrea Scalco[1], Andrea Ceschi[1], Riccardo Sartori[1], and Enrico Rubaltelli[2]

[1] University of Verona, Department of Philosophy, Education and Psychology, Verona, Italy
{andrea.scalco,andrea.ceschi,riccardo.sartori}@univr.it
[2] University of Padova, Department of Developmental Psychology and Socialisation, Padova, Italy
enrico.rubaltelli@unipd.it

Abstract. In the present study we developed a simulation where agents play repeatedly the ultimatum game with the aim of exploring their earnings for several thresholds of willingness to accept proposals. At the same time, the scope is also to provide a simple and easily understandable simulation of the ultimatum game. Particularly, the simulation generates two kinds of agents, whose proposals are generated accordingly to their selfish or selfless behavior; subsequently, agents compete in order to increase their wealth playing the ultimatum game with a random-stranger matching. The trend emerged by simulation charts shows how, even when altruistic agents bid higher proposals than those following selfish behaviors, the average mean cash earned is higher for the former agents than the latter. A second fact is that, looking at the system as a whole, altruistic punishment leads to a reduction of the resources exploited by the agents. Finally, we introduced the psychological construct of trait emotional intelligence, briefly discussing the value of its implementation into computational simulations.

Keywords: Ultimatum game, Agent-based model, Altruistic behavior, Trait emotional intelligence.

1 The Dilemma behind the Ultimatum Game

The ultimatum game (UG) is a well-known sum-zero game composed of a single bargain, illustrated for the first time by Güth, Schmittberger and Schwarze [1]. It represents an ideal situation of economic interaction between two individuals. The rules of the ultimatum game are quite elementary and can be easily summarized as follows. The first player is endowed with a certain amount of money and must decide how to split it with a second player. The latter can accept or refuse the division proposed: if he/she accepts, both players will earn money according to the proposal; if he/she refuses, neither of them will gain anything. According to the idea of rationality proposed by the economic theory, it would be inevitable that the proposer had to split the money in order to chase the maximum profit: therefore, he should give not more than the possible minimum to the responder. Accordingly, if the responder follows the

J. Bajo et al. (eds.), *Trends in Prac. Appl. of Agents, Multi-Agent Sys. and Sustainability*,
Advances in Intelligent Systems and Computing 372, DOI: 10.1007/978-3-319-19629-9_22

normative approach, he should accept without doubts, because the minimum offer is certainly better than nothing [2]. In spite of these considerations, this is not what happens in real world. In fact, as reported by Camerer [3] (who examined over 30 studies concerning the ultimatum game), the most common strategy followed by the proposers is more or less based on equity; similarly, even the responders show an irrational behavior, given that the proposals are usually rejected if they are perceived as unfair. This last act has been defined as an altruistic punishment [4]: by refusing the proposal, the responder punishes the proposer deleting all his chances to gain the money, although this means to sacrifice his/her own profit. In other words, the decision of the proposer, as well as the receiver, is affected by the individual emotional experience during the game: an idea that has been corroborated even by neurosciences [5, 6]. Following this consideration, we believe that a contribution to the study of altruistic behaviors could derive from the study of emotional intelligence (EI) within interaction contexts. This latter represents a psychological construct that has been widely discussed during the last decade due to the emergence of numerous conflicting theories: [7] have extensively faced the concept, and its associated theoretical issues, through several domains. However, we can reconcile the theories about emotional intelligence when we state that it can be identified like the ability to label and control own feelings, as well as understand and influence others. Consequently, it is conceived as a superior dimension strictly associated with personal experience of emotional states. As emotional intelligence is positively associated with the social ability and empathy [8], we can suppose that proposers with a high level of EI are able to put themselves in the receiver' shoes: in this way, they are able to think strategically to the interaction.

At any rate, it is clear that, in just a single round of the ultimatum game, the first player must decide carefully his/her proposal in the direction of not trigger off the refusal of the second one: so, in order to increase the chance to win, he/she should move towards altruistic proposals and give up part of his/her own profit. Nevertheless, when multiple rounds of UG are played a question arises: does altruistic behavior really lead to worse economic performance than a selfish act? Considering this, in our study we developed a simple simulation where two different types of agents (selfish and altruistic) play repeatedly (and with a random-stranger matching) the ultimatum game. This was done in order to study their earnings for several thresholds of willingness to accept proposals. At the same time, the scope is also to provide a simple and easily understandable simulation of ideal negotiation providing its source code: in this way, we hope to encourage psychologists, as well as other social scientists, to begin to appreciate computational simulations to study altruistic behavior, prosocial acts, and decision-making processes related to emotions. Therefore, the code of the simulation has been uploaded and can be retrieved inside the Model Library page of the OpenABM Consortium site. [9]

2 Computational Simulation and ABMs

The major value of computational simulation lies on their ability to reproduce individual and social behavior through a computer language and a dedicated software. Particularly, agent-based models (ABMs) are "a type of computer simulation that is used to study how micro-level processes affect the macro-level outcomes" [10, p. 488]. This approach is the most useful to be applied when it is not possible to figure out the outcome of a simulation by an analytical way comprised of differential equations [11]. In this way, Axelrod [12] has defined simulation as "a third way of doing science". It starts from deduction, defining some assumptions, but it does not prove theorems. On the contrary, it generates data that are analyzed inductively, but these are clearly generated from previous assumptions, instead of being gathered from real world.

Again, Axelrod [12] listed a variety of purposes of simulations. They are useful to discover and understand the effects of hypothetical social mechanisms. They can be used to perform tasks, as typically happens in the field of artificial intelligence. They are also used to train people, helping them to figure out solutions in complex environments. Finally, ABMs can be useful to predict outcomes from several inputs: this is the case of our study, where we have explored the results of several random interactions of agents.

We believe that agent-based modeling could represents an intuitive and agile methodology compared to analytical methods. To start working on computational simulations we would suggest reading the guide provided by Axelrod and Tesfatsion [13].

3 Simulating Iterated Ultimatum Games with Random Players

To build the simulation we made use of NetLogo v.5.1.0, developed by Wilensky as a "multi-agent programming language and modelling environment" [14]. NetLogo is very common in simulation field thanks to its easy programming language [15,16]. Thus, coherently with the purposes of the paper, we decided to use this platform to provide a clear and simple starting point for other psychologists interested in studying ideal situations of interaction and behaviors related to the emotional area. Furthermore, so far a model to simulate iterated ultimatum games has not been added in the Model Library of NetLogo.

The agents implemented into the simulation are divided into two groups according to their behavior, and following the observations of Camerer [3]. The amount of money to divide inside the ultimatum game is fixed at 100.00 €. The first group acts in a selfish way, that is, their proposals are always generated randomly under the 50% of the amount of money to split (more in detail, their proposals range from 35.00€ to 45.00€). Conversely, the second group behaves in the opposite way: indeed, altruistic agents bid higher, even over the half of initial amount of money (particularly, proposals randomly range from 45.00€ to 55.00€). The conducted simulations have not taken into account learning behaviors based on players experience or the changing of the initial amount of money to split, given that, as reported by Camerer [3], the studies which exploited the UG showed only small effects for these factors.

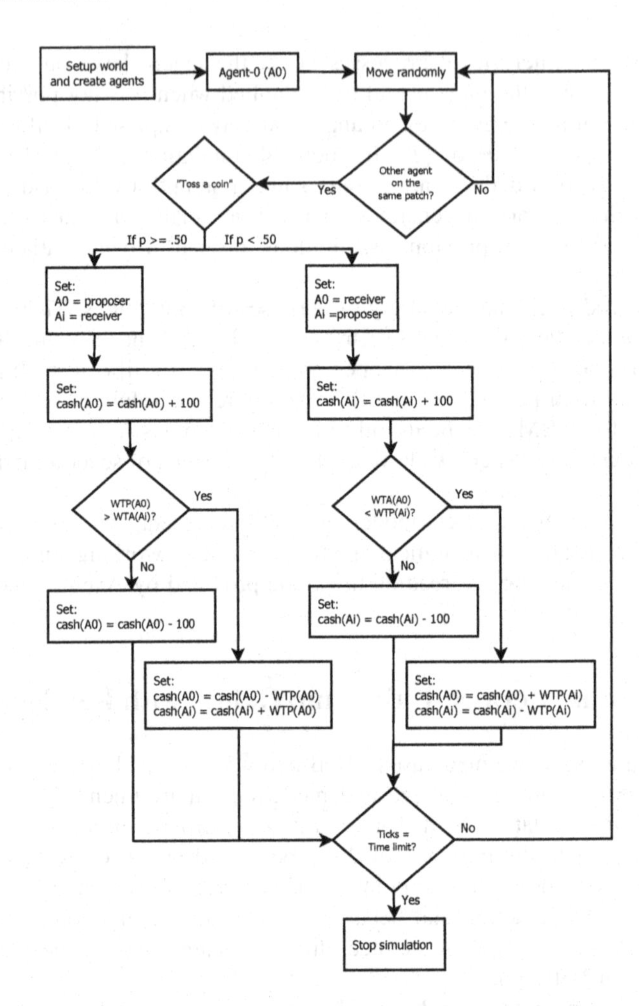

Fig. 1. The algorithm shows the steps computed by a single agent (Agent 0 or A0) moving around the virtual world and encountering accidentally another agent (Agent i or Ai)

Specifically, the programmed simulation can create N agents (precisely, from 100 to 1.000 by 100) which are able to autonomously move randomly inside the virtual world. Each agent is endowed with: a random willingness to pay (WTP), accordingly to the behavior of the agent; a minimum of money that the agent would accept in the ultimatum game (willingness to accept; WTA). The latter is fixed equally for all agents and it is alterable by the user through the interface tab of the program. In this way, it is possible to run different simulations changing the number of agents in the simulation and their willingness to accept proposals. The interface tab allows also changing the percentage of altruistic agents to be generated (from 0% to 100%). The flow chart in Fig. 1 represents the steps of a single agent moving in the artificial world. Briefly, the agent can move randomly at each step of the simulation and, when it finds someone on its same position, it plays with it the ultimatum game. If it wins, the earned cash is stored into the agent's virtual wallet; if it loses, the agent exits from the current game leaving its economic resources untouched. After it has played, it goes back to move randomly. Thus, in this way the virtual game follows the rules of a single-shot UG, as previously reported in the introduction.

By analogy, it is possible to conceive a situation like this in which two agents meet: a person walking accidentally finds 100.00 € on the street, but at the same time, another person arrives at the same place and, similarly, finds the money. Thus, in order to decide who needs to take the banknote, they toss a coin. Nonetheless, they agree on the fact that who will take the money, will also have to split it with the other one, following what is reasonable for him. If the responder accepts the amount of cash proposed, they will both earn the cash, otherwise they agree to throw it away, losing nothing.

The code, which can be downloaded from the Model Library of the ABM Consortium site [9], should appear clear and easy to understand even to newcomers of simulation field after they will have gained just some experience with the NetLogo platform. We hope it could be also intuitive and be the starting point for psychologists who want to begin working on computational simulations. The agents' behavior could be retrieved inside the procedure called "initBehav": this is the part of the program that (more than others) could be rewritten in order to provide different and complex behaviors to agents, leaving unaltered the "engine" of the simulation which works on the iteration of the ultimatum game.

4 Data Analysis

The simulation we run took into account different scenarios based on the number of agents in the world and several values indicating the willingness to accept proposals. Precisely, the number of agents considered ranges from 100 to 1.000 by 100, whereas WTA has been analyzed (for each of the previous states) starting from 30.00€ to 50.00€ by 1. For the analysis, we decided to limit the simulation time at 750 cycles. For each population of agents and each WTA considered, we applied 10 runs in order to obtain an average result, given the randomness nature of simulations. Consequently, for each result of these ten runs the average cash won by the agents acting selfishly and selflessly has been calculated. In this way, it is possible to observe the average cash earned by the two groups based on different values indicating the willingness to accept proposals (Fig. 2). Furthermore, we tried to change the percentage of agents endowed with an altruistic behavior, but we did not observe any significant changes in the final trends extracted from the charts comparing the two groups. Finally, the differences that can be observed between trends among conditions are mainly due to the random generation of the agents' willingness to pay.

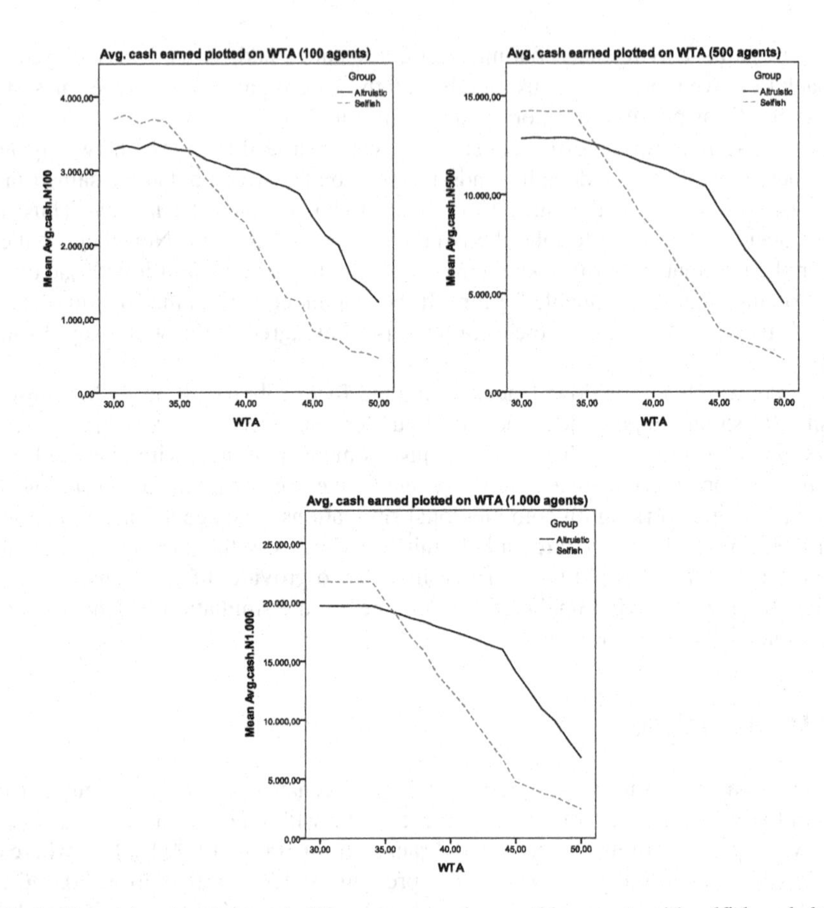

Fig. 2. Charts showing the trend of the average cash earned by agents with selfish and altruistic behavior for different numbers of agents (100; 500; 1.000) with WTA ranging from 30,00€ to 50,00€. Each value associated to a single WTA is the average result of ten runs of the simulation.

5 Results and Discussion

The program reproduces an ideal economic condition where a variety of agents, distinguished for their altruistic or selfish behavior, must randomly compete among them in order to increase their wealth (i.e. raise their portfolio). The model demonstrates the potential of simulations, that allow to understand the average outcome in terms of earned money through the repeated and random interaction of the agents, without the necessity to conduct expensive studies (it would be quite impossible, or at least very difficult, to reproduce this kind of research inside a laboratory with real people) or as an intuitive and agile alternative to analytical methods.

The conducted simulation shows how, even when altruistic agents bid higher proposals than those following selfish behaviors, if we look at the whole system, the average mean cash earned is better for the former agents than the latter. Nonetheless, this is true just when the willingness to pay is set above a certain threshold. Indeed, if WTA falls below the threshold of 35-36% of the initial amount of money, the best average performance is achieved by selfish players. The charts in Fig. 2 illustrate these results, showing the average cash earned by the two different kinds of agents plotted on several WTA for different populations. It is notable that the average WTA observed through several experimental studies usually swing near this threshold: indeed, Camerer [3] reported how the proposal is rejected half of the time when under the 20% or so and quite always accepted when at least equal to 40-50% of the initial amount of cash.

Furthermore, an interesting fact is that, looking at the system as a whole, the rejection of the proposals with the consequences of losing money (also known as altruistic punishment), when largely applied, leads to a reduction of the resources exploited by the agents (we should remind that the amount of money to divide is indeed "donated" to players). In fact, it is possible to observe in Fig. 2 that, when WTA exceeds the threshold of 35-36%, just as the trend of earnings of each group overturns, both the gain of selfish and altruistic agents significantly decrease.

Finally, we would like to use this opportunity to state how a significant contribution to the understanding of the apparent irrational behavior that leads people to reject offers, even sacrificing their own profit, could stem by the introduction of the concept of emotional intelligence into computational models. Emotional intelligence is conceived as a personality trait and could be identified as an aggregation of emotional self-perceptions [17]. Particularly, this psychological construct concerns individual differences on the perception, and thus management, of the emotional sphere. Petrides' work [18] reports several studies to appreciate how EI can represent a worthy predictor of prosocial and antisocial behavior, of emotion regulation, of mood, as well as affective decision-making. A good discussion of the relationship between EI and decision-making and its importance regarding the study of individual differences on choices has been provided by Sevdalis, Petrides and Harvey [19]. We think that its implementation into a computational model could advance the development of heterogeneous population of virtual agents.

Following this way, our successive work [20] will be based on the simulation and code program here presented, but agents' behavior will be rewritten accordingly to the results of an experimental study concerning the relation between EI and WTP in the ultimatum game. The data were gathered by participants who responded to a survey measuring the level of EI, their money attitude and in which they were asked to play the ultimatum game three times varying the initial amount of money. In this study, we observed a positive correlation between EI and willingness to pay. These findings will act as the base to implement the complex behavior of the agents: particularly, the EI and attitude to money will be the ground on which agents will make their proposals.

References

1. Güth, W., Schmittberger, R., Schwarze, B.: An experimental analysis of ultimatum bargaining. J. Econ. Behav. Organ. 3, 367–388 (1982)
2. Haselhuhn, M.P., Mellers, B.A.: Emotions and cooperation in economic games. Cogn. Brain Res. 23, 24–33 (2005)

3. Camerer, C.F.: Behavioral game theory: Experiments in strategic interaction. Russel Sage Fondation, New York (2003)
4. Fehr, E., Gächter, S.: Altruistic punishment in humans. Nature 415, 137–140 (2002)
5. Zak, P.J., Stanton, A.A., Ahmadi, S.: Oxytocin increases generosity in humans. PLoS One 2, e1128 (2007)
6. Sanfey, A.G., Rilling, J.K., Aronson, J.A., Nystrom, L.E., Cohen, J.D.: The neural basis of economic decision-making in the ultimatum game. Science 300, 1755–1758 (2003)
7. Stough, C., Saklofske, D.H., Parker, J.D.A.: Assessing emotional intelligence: Theory, research, and applications. Springer, New York (2009)
8. Schutte, N.S., Malouff, J., Bobik, C., Coston, T.D., Greeson, C., Jedlicka, C., Rhodes, E., Wendorf, G.: Emotional intelligence and interpersonal relations. J. Soc. Psychol. 141, 523–536 (2001)
9. Model Library of the OpenABM Consortium, https://www.openabm.org/models. The discussed model can be retrieved at https://www.openabm.org/model/4552/version/2
10. Hughes, H.P.N., Clegg, C.W., Robinson, M.A., Crowder, R.M.: Agent-based modelling and simulation: The potential contribution to organizational psychology. J. Occup. Organ. Psychol. 85, 487–502 (2012)
11. Terna, P., Boero, R., Morini, M., Sonnessa, M.: Modelli per la complessità. La simulazione ad agenti in economia [Models for complexity. Agent-based simulation in economy]. Il Mulino, Bologna, Italy (2006)
12. Axelrod, R.: Advancing the art of simulation in social science. In: Rennard, J.-P. (ed.) Handbook of Research on Nature Inspired Computing for Economy and Management, pp. 1–13. Idea Group, Hersey (2005)
13. Axelrod, R., Tesfatsion, L.: A Guide for Newcomers to Agent-Based Modeling in the Social Science. In: Tesfatsion, L., Judd, K.L. (eds.) Handbook of Computational Economics. Agent-Based Computational Economics, vol. 2, pp. 1–13. Handbooks in Economics Series, Amsterdam (2005)
14. Wilensky, U.: NetLogo. Center for Connected Learning and Computer-Based Modeling. Northwestern University, Evanston (1999), http://ccl.northwestern.edu/netlogo/
15. Gilbert, N., Troitzsch, K.G.: Simulation for the social scientist, 2nd edn. Open University Press, Maidenhead (2005)
16. Railsback, S.F., Lytinen, S.L., Jackson, S.K.: Agent-based simulation platforms: Review and development recommendations. Simulation 82, 609–623 (2006)
17. Petrides, K.V.: Trait emotional intelligence theory. Ind. Organ. Psychol. 3, 136–139 (2010)
18. Petrides, K.V.: Ability and trait emotional intelligence. In: Chamorro-Premuzie, T., von Stumm, S., Furnham, A. (eds.) The Wiley-Blackwell Handbook of Individual Differences, pp. 656–678. Blackwell Publishing Ltd., Hoboken (2011)
19. Sevdalis, N., Petrides, K.V., Harvey, N.: Trait emotional intelligence and decision-related emotions. Pers. Individ. Dif. 42, 1347–1358 (2007)
20. Scalco, A., Rubaltelli, E., Sartori, R., Ceschi, A.: Does altruistic behaviour lead to better economic profits than self-interested behaviours? A counterintuitive response from an ABM (working paper)

Compassion and Prosocial Behavior. Is it Possible to Simulate them Virtually?

Andrea Ceschi[1], Andrea Scalco[1], Stephan Dickert[2,3], and Riccardo Sartori[1]

[1] University of Verona, Department of Philosophy, Education and Psychology, Verona, Italy
{andrea.ceschi,andrea.scalco,riccardo.sartori}@univr.it
[2] WU Vienna University of Economics and Business
[3] Linköping University, Linköping, Sweden
stephan.dickert@wu.ac.at

Abstract. In the field of artificial intelligence, a question dealing with computer and cognitive science is arising and becoming more and more crucial: Can we design agents so sophisticated that they are capable of mimicking emotional behaviors in general as well as specific emotions like compassion or empathy? Despite the production of different computational models, their integration with cognitive and psychological theories remains a central problem. Reasons are both methodological and theoretical. Primarily, it is difficult to quantify the impact of such factors as individual differences, inclinations and personality traits. In addition, *Agent-Based Models* (ABMs) often use linear dynamics, even in describing emotions, without considering the basis of psychophysics. Bearing in mind this and focusing on compassion as a particular emotion, the paper aims to present a "Decalogue" for those interested in designing agents capable of mimicking human emotional behaviors. In the paper, compassion will be translated as prosocial behavior.

Keywords: Emotional behaviors, Compassion, Prosocial Behavior, Psychophysics.

1 Advances and Limits in Computing and Reproducing Emotional Behaviors

Emotions are considered important for computational models because they can provide both an explanation for agents' adaptive behaviors (Anderson et al., 2004; Scheutz & Sloman, 2001) and ideas to facilitate social cooperation in real life (Gratch & Marcella, 2001; Keltner & Haidt, 1999). Gratch and Marsella (2004) consider the development of computational models of emotions as a core research focus that will facilitate both the interpretation of and the influence on human behavior. Emotion is essential for thought as much as thought is essential for emotion (Lazarus, 1982). That is why it is important to consider them together. Lin, Spraragen, Blythe, and Zyda (2011) describe three main computational models that show different aspects of the relationship between emotion and cognition:

J. Bajo et al. (eds.), *Trends in Prac. Appl. of Agents, Multi-Agent Sys. and Sustainability*,
Advances in Intelligent Systems and Computing 372, DOI: 10.1007/978-3-319-19629-9_23

1. *EMA* (Marsella & Gratch, 2009): Mostly interested in explaining the dynamics of emotions through a sequence of events, it focuses on appraisal and how this leads first to emotions and then to coping. The model shows how building not-confirmed expectations could lead to very specific emotions (delusion, frustration, etc.).
2. *Soar-Emote* (Marinier III, Laird, & Lewis, 2009): It shows how different appraisals influence the notion of intensity. In particular, authors propose to use appraisal to calculate both arousal and valence in various models.
3. *WASABI* (Becker-Asano & Ishiguro, 2009): It is supposed to be one of the most general models of emotions built to simulate affective agents. The model uses primary and secondary emotions. The first ones are more instantaneous while the second ones are more complex and arise from reasoning.

Since then, much progress has been made both in the field of computer science and in the study of emotions. Despite the production of different computational models, their integration with solid psychological and cognitive theories remains a central problem . Reasons are both methodological and theoretical. Primarily, the lack of interdisciplinarity makes it difficult to adapt theories to classical models based on agents without simplifying the factors present in them. Secondly, it is difficult to quantify the impact of such factors as individual differences, inclinations and personality traits (Sartori, 2010; Sartori & Pasini, 2007). Thirdly, *Agent-Based Models* (ABMs) often use linear dynamics, even in emotions, without considering the basis of psychophysics.

Despite the limitations presented so far, it is important to state that ABMs represent a true revolution for psychology and cognitive science, probably the first instrument able to dynamically reproduce or mimic emotions connected to behaviors. All this considered, the paper aims to present a "Decalogue" for those interested in designing agents capable of mimicking human emotional behaviors. In particular, the present contribution will try to give some useful suggestions in order to answer the three limits mentioned above by presenting a case study related to compassion and prosocial behavior.

2 Using Cognitive and Psychological Studies for Better Simulations

In order to design agents with appropriate characteristics, interdisciplinary methods are crucial. In particular, it is worth considering those theories developed by psychology and cognitive science that outline perspectives that should be considered in designing agents. In the specific case of compassion and prosocial behavior, social scientists and cognitivists have developed several theories to explain them. Batson and Oleson (1991), for example, believe that the main goal of prosocial behavior is to make others feel better, and, in this regard, empathy assumes a leading role. Empathy is a social-affective dimension that can be developed by training experiences and is at the basis of interaction and relationships (Cunico et al., 2012). Other theories are more "egoistic" than this and suggest that, when deciding to give help, our benefits are more important than others' benefits. Andreoni (1989) states that, when we help someone, we experience a sort of "warm-glow" feeling inside and this would be the main motivation for altruism. On the other hand, Cialdini et al. (1987) think that we give help particularly when the suffering of others leads to negative feelings that we want to alleviate. Most scientists think that once we are able to identify the main motivators we can understand and predict prosocial behavior more easily. However, the question is whether experimental studies are able to respond to a huge number of interactions of different types while they are happening between people who give help and people who need help. What happens, for example, if, after you have just helped someone, you meet a person that you cannot help or even two people in need, one after the other, which cannot be helped? How does compassion change over time in a system where there are people who cannot be helped? For social psychologists and cognitive scientists, these questions appear to be at the same time extremely interesting and difficult to answer by simply relying on classical experimental approach. For this reason, the use of ABMs is becoming precious, especially if these models are computed by starting from empirical theories.

3 Better to Start from SEMs. The Importance of the Empirical Data

It is well known that the development of ABM algorithms is the most fragile aspect of the simulation analysis. In the specific case of ours, in order to design social networks, the crucial preference is the identification of an amount of agent's attributes significant for prosocial behavior. The theories we have dealt with before are able to help. Agents' attributes span from basic demographic attributes to more specific and psychological features (Ceschi, Sartori, & Rubaltelli, 2014). These attributes have the greatest impact and therefore it is important to consider them for the aims of the analysis. For this reason, it is recommended to start from some empirical models, such as *Structural Equation Models* (SEMs), which are widespread in social and psychological science and they are used to represent human behaviors. They consist of equations that link latent variables and their indicators by using exogenous causes. A remarkable benefit of this framework is that correlations of observed indicators are clearly made as arising out of subjacent factors that are accountable for the results (Krishnakumar & Ballon, 2008). Apart from this, results from SEMs have never been dynamically linked to individual differences. This is a limit of this quantitative method. Essentially, individual differences are characterized as a set that makes individuals particular, according to their inclinations, capabilities and outcomes. Starting from here, ABMs can help to represent, in a natural and dynamic way, several scales of analysis and the importance of structures at different levels, none of which is easy to accomplish with other modeling techniques (Gilbert & Terna, 2000).

In the current paper, the aim of the analysis is to present a model able to simulate a number of characteristics that have been scaled, modeled and assigned with probability distributions to simulate prosocial behavior. Usually, the purpose of this stochastic effort is to endow agents with a 'personality'. For this reason, SEMs on prosocial behaviors (Maner & Gailliot, 2007; Ranganathan & Henley, 2008) have been considered as determinant factors of prosocial behavior itself.

4 How to Compute Compassion over Time. A Solution from Psychophysics

Commonly, the good feeling you get when you help someone in need drops over time following the psychophysical function of human sensitivity (Weber, 1834). While this sounds normal in psychophysical terms (imagine the difference in sensation when you turn on the light in a dark room and when you do the same thing in an already lighted space), how normal does that sound when we evaluate somebody's life differently just because they are the first or the nth needing help?

On the other hand, the greater the number of people you cannot help, the lower is the good feeling of you experience about those you can help. Fetherstonhaugh, Slovic, Johnson and Friedrich (1997) found that people have a diminishing sensitivity to the value of life over time, which they called *Psychophysical Numbing* (Figure 1). Västfjäll and Slovic (2011) found that participants reported a decrease in feeling good about giving help during the increase of the number of people that could not be helped. The phenomenon has been called the *Pseudo-Inefficacy effect* (Västfjäll & Slovic, 2011) and it is a further example that people not always respect the utilitarianism norms (Dickert, Västfjäll, Kleber, & Slovic (2012).

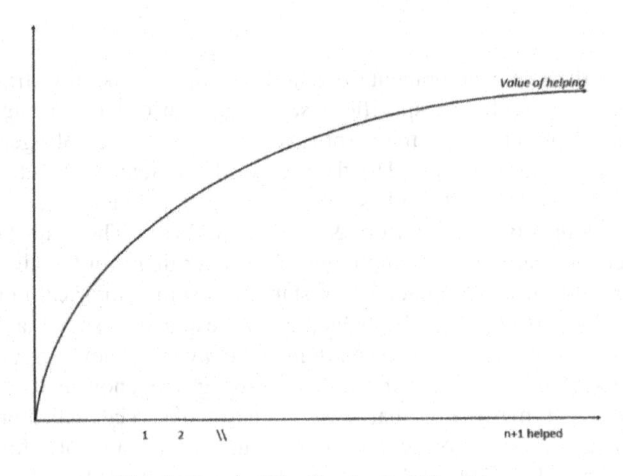

Fig. 1. Trend of Psychophysical Numbing

5 A Proposal for a Realistic Simulation of Compassion and Prosocial Behavior

Naturally, factors that motivate helping and factors that demotivate helping are connected. It is therefore fundamental to study them in their interactions with altruistic behavior. Designing agents capable of reproducing compassion and prosocial behavior requires different algorithms that must be integrated so that they can work together and take into considerations the points and theories previously seen.

In order to create a model capable of recreating compassion, it is important to consider that "compassionate agents" will be influenced by both agents that can be helped and cannot be helped. "Compassionate agents" should be demotivated if they meet agents who cannot be helped. On the other hand, they should be motivated after helping others. The model should include 3 variables which play an important role in prosocial behavior: 1. *Compassion* (*C*), 2. *Demotivation* (*D*) and 3. *Altruism* (*A*).

Let's suppose that *A* is the need to help that individuals feel in certain situations and that they will be repaid with *C*. This mechanism happens when agents can help other agents. If the helper cannot find someone to help, functions invert their trends (*C* starts decreasing, while *A* starts increasing). *D* decrements *A* and *C* levels when the agent meets others that cannot be helped. In order to compute mathematically these trends (Figure 2), we used several logarithmic and decay functions based on coefficients extracted from SEMs on prosocial behaviors (Maner & Gailliot, 2007; Ranganathan & Henley, 2008).

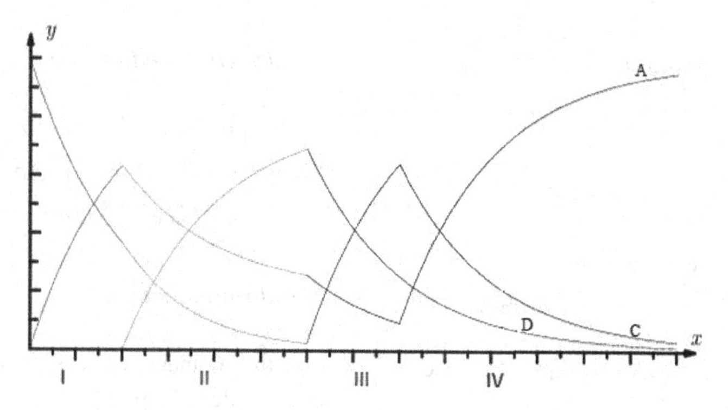

Fig. 2. Trends of Compassion, Demotivation and Altruism, where *X* is the standardized level of *C*, *D* and *A* or of the feeling observed, and *Y* the time

In phase I, an agent has the possibility to help agents in all the time units. Consequently, its level of *A* increases and its level of *C* decreases. In phase II, the helper starts meeting agents that cannot be helped. Its levels of *C* and *A* go down while *D* increases. In phase III, the helper finds agents to help, and, therefore, *D* stops amounting and starts decreasing. *C* increases, *A* decreases. In phase IV, the helper is no more able to find other agents. Consequently, *C* decreases and *A* increases, while *D* continues to decrease. Next section will suggest a mathematical model for the example here reported (Table 1).

Table 1. Definitions and computation of the variables of the present example

Name	Meaning
t	the time variable
C	the Compassion level
D	the Demotivation level
A	the Altruism level
C_0	the maximum value of Compassion
m	the states of agents are: a. when the agent does not meet others; b. when the agent meets an agent in need of help; c. when the agent meets an agent in need of help, but is not able to help it.
λ	the constant value of the Demotivation factor which influences the Compassion
μ	the constant value of the Demotivation factor

The compassion function

C function of phase I

$$C(t)=\begin{cases} C_0 e^{-t} \\ C_0(1-e^{-t}) \end{cases}$$

Equations are used to compute the t_1 and t_0 respectively.

C function of phase II

$$C(t)=\begin{cases} C_0 e^{-(t-(t_3-t_0))} & \mid m=a \\ C_0(1-e^{-(t-(t_2-t_1))}) & \mid m=b \end{cases}$$

Where t is the current time.
t_0 is the actual time when Equation reaches the value $C(t_3)$
t_1 is the actual time when Equation reaches the value $C(t_2)$.
t_2 is the moment when the agent does not find people to help.
t_3 is the moment the agent find people he can help.

C function of phase III

$$t_1 = ln(C_0 / C_{t_3})$$

$$t_0 = ln(C_0 / (C_0 - C_{t_2}))$$

C function of phase IV

$$C(t)=\begin{cases} C_0 e^{-(t-(t_3-t_0))} & \mid m=a \\ C_0(1-e^{-(t-(t_2-t_1))}) & \mid m=b \\ 1-\lambda^{t-t_1} C_0 e^{-t} & \mid m=c \end{cases}$$

The demotivation function

$$D(t)=\begin{cases} 0 & \mid m=a \\ C_0 e^{-(t-(t_2-t_1))} & \mid m=b \\ C_0(1-e^{-(t-t_{III})}) & \mid m=c \end{cases}$$

The altruism function

t_2 is the moment when the agent does not find people to help.

$$A(t)=\begin{cases} C_0(1-e^{-(t-(t_3-t_0))}) & \mid m=a \\ C_0(1-e^{-(t-(t_2-t_1))}) & \mid m=b \\ C_0(1-e^{-(t-(t_3-t_0))}) - C_0 e^{-(t-(t_3-t_0))} & \mid m=c \end{cases}$$

Where t is the latest moment when the agent does not find people to help.

6 Conclusion

As Gratch and Marsella (2004) claim, including elements related to emotions in ABMs can facilitate both the interpretation of and the influence on human behavior. In this model, we used compassion as a predictor of the helping behavior, following a method that takes into account solid psychological theories and experimental data.

The model, here presented just for the mathematical part, is finding application in a simulation that aims to recreate agents capable of reproducing compassion and the feeling of inefficacy that can be found often in charity giving issues, where the help needed is always bigger than the material resources that others can offer. As well, by using these algorithms, we aim to simulate contexts to manage natural disasters, epidemic situations, medical problems and several other events where helping behavior is crucial.

As for this basic model, we only included a simple relation between agents in need. On the other hand, and quoting the famous saying by Korzybski, *the map is not the territory*. In real life, there can be many other factors that would be worth being implemented in this relation. Demotivation and motivation to help can be mediated or moderated by other factors such as mood, previous experiences (not everyone feels the same demotivation when they learn about individuals they cannot help), coping strategies, etc. On the other side, the actual help might be also influenced by the characteristics of the individuals in need, such as gender, race, physical conditions and others. Several other characteristics could be implemented into the mathematical model in order to run even more realistic simulations regarding prosocial behavior.

References

1. Anderson, J.R., Bothell, D., Byrne, M.D., Douglass, S., Lebiere, C., Qin, Y.: An integrated theory of the mind. Psychological Review 111(4), 1036–1060 (2004)
2. Andreoni, J.: Giving with impure altruism: Applications to charity and Ricardian equivalence. Journal of Political Economy 97, 1447–1458 (1989)
3. Batson, C.D., Oleson, K.C.: Current status of the empathy-altruism hypothesis. In: Clark, M.S. (ed.) Review of Personality and Social Psychology. Prosocial behavior, vol. 12, pp. 62–85. Sage, Newbury Park (1991)
4. Becker-Asano, C., Ishiguro, H.: Laughter in social robotics—no laughing matter. Paper Presented at the Intl Workshop on Social Intelligence Design (SID 2009) (2009)
5. Ceschi, A., Rubaltelli, E., Sartori, R.: Designing a Homo Psychologicus More Psychologicus. Empirical Results on Value Perception in Support to a New Theoretical Organizational-Economic Agent Based Model. In: Omatu, S., Bersini, H., Corchado Rodríguez, J.M., González, S.R., Pawlewski, P., Bucciarelli, E. (eds.) Distributed Computing and Artificial Intelligence, 11th International Conference. AISC, vol. 290, pp. 71–78. Springer, Heidelberg (2014)
6. Cialdini, R.B., Schaller, M., Houlihan, D., Arps, K., Fultz, J., Beaman, A.L.: Empathy-based helping: Is it selflessly or selfishly motivated? Journal of Personality and Social Psychology 52, 749–758 (1987)
7. Cunico, L., Sartori, R., Marognolli, O., Meneghini, M.: Developing empathy in nursing students: a cohort longitudinal study. Journal of Clinical Nursing, 2016–2025 (2012)
8. Dickert, S., Västfjäll, D., Kleber, J., Slovic, P.: Valuations of human lives: normative expectations and psychological mechanisms of (ir) rationality. Synthese 189, 95–105 (2012)
9. Fetherstonhaugh, D., Slovic, P., Johnson, S.M., Friedrich, J.: Insensitivity to the value of human life: A study of psychophysical numbing. Journal of Risk and Uncertainty 14, 283–300 (1997)
10. Gilbert, N., Terna, P.: How to build and use agent-based models in social science. Mind & Society 1(1), 57–72 (2000)
11. Gratch, J., Marcella, S.: Modeling the Interplay of Emotions and Plans in Multi-Agent simulations. Paper presented at the Proceedings of the Cognitive Science Society (2001)
12. Keltner, D., Haidt, J.: Social functions of emotions at four levels of analysis. Cognition & Emotion 13(5), 505–521 (1999)
13. Krishnakumar, J., Ballon, P.: Estimating basic capabilities: A structural equation model applied to Bolivia. World Development 36(6), 992–1010 (2008)
14. Lazarus, R.S.: Thoughts on the relations between emotion and cognition. American Psychologist 46, 352–367 (1982)

15. Lin, J., Spraragen, M., Blythe, J., Zyda, M.: Emocog: Computational integration of emotion and cognitive architecture. Paper presented at the Proceedings of the Twenty-Fourth FLAIRS Conference (2011)
16. Maner, J.K., Gailliot, M.T.: Altruism and egoism: Prosocial motivations for helping depend on relationship context. European Journal of Social Psychology 37(2), 347–358 (2007)
17. Marinier III, R.P., Laird, J.E., Lewis, R.L.: A computational unification of cognitive behavior and emotion. Cognitive Systems Research 10(1), 48–69 (2009)
18. Ranganathan, S.K., Henley, W.H.: Determinants of charitable donation intentions: a structural equation model. International Journal of Nonprofit and Voluntary Sector Marketing 13(1), 1–11 (2008)
19. Sartori, R.: Face validity in personality tests: psychometric instruments and projective techniques in comparison. Quality & Quantity 49, 749–759 (2010)
20. Sartori, R., Pasini, M.: Quality and Quantity in Test Validity: How can we be Sure that Psychological Tests Measure what they have to? Quality & Quantity 41, 359–374 (2007)
21. Scheutz, M., Sloman, A.: Affect and agent control: Experiments with simple affective states. In: Intelligent Agent Technology: Research and Development, pp. 200–209 (2001)
22. Västfjäll, D., Slovic, P.: Pseudo-inefficacy: When awareness of those we cannot help demotivates us from aiding those we can help. Manuscript in preparation (2011)
23. Weber, E.H.: De pulsu, resorptione, auditu et tactu. Koehler, Leipzig (1834)

Intelligent Tutoring System, Based on Video E-learning, for Teaching Artificial Intelligence

Antonio Bailón, Waldo Fajardo, and Miguel Molina-Solana

Dept. Computer Science and AI. Universidad de Granada, Granada, Spain
{bailon,aragorn,miguelmolina}@ugr.es

Abstract. In the last few years, distant learning is gaining traction as a valid teaching approach taking advantage of the Internet and current multimedia capabilities. Even though thousands of students are enrolling to *Massive Open Online Courses*, there is still a lack of proper educative programs who account for the individual characteristics of the students. In particular, most e-learning courses tend to be mere repositories of contents, very teacher-centric and lacking the necessary individual personalisation to account for each student's needs, expectations and paces. In this work, we propose and describe an Intelligent Tutoring System that enables the automatic adaptation of the contents of the course to the particular learners. The systems was tested with a group of students with very positive direct and indirect results.

1 Introduction

Even though many tools for distant learning exist nowadays, they tend to be mere information repositories simulating textbooks. Only sometimes they are effectively complemented with communication tools such as forums, emails and online meetings that enables the communication between tutors and participants. More recently, a new model of platforms has emerged in which most of their contents are presented as videos, with almost no importance for written materials [1]. The concept of community within a course, to solve doubts or comment on the contents, is also gaining major traction. Additionally, in those courses, learners sometimes also act as course tutors, by grading peers works and solving questions, freeing the real tutors of these tasks.

However, in both the aforementioned approaches, the role of the tutor as the individual who guides and personalises the students learning according to their effort and capabilities does not exist or is very diminished. Therefore, the learning process becomes a depersonalised one, in which all students have the same work plan, with no consideration to their personal particularities. This is an undesirable situation, as it gives no attention to the more needed students, and can make the learning process very boring for the most qualified students.

The present work introduces the design, development and implementation of an e-learning tool based on video-teaching, that incorporates an *Intelligent Tutoring System (ITS)*. Additionally, we show its application to a particular real case, demonstrating how the inclusion of the ITS enables the personalisation of

J. Bajo et al. (eds.), *Trends in Prac. Appl. of Agents, Multi-Agent Sys. and Sustainability*, Advances in Intelligent Systems and Computing 372, DOI: 10.1007/978-3-319-19629-9_24

the learning process and avoids the common e-learning problems related to the un-adaptation of the contents to the learners. The system we present reproduces the teachers behaviour on its role of decision maker based on student supervision.

In the work, we describe previous tools, along with the characteristics, the teaching methodology, the architecture, the implementation and the additional functionality of our proposal of intelligent tutoring system, and its application to the real case in a course at our university. The paper concludes with a reflection on the aforementioned points.

2 Background

Traditionally, the lecture based in a rigid schema in which the teacher has an active role and the students a passive one, has been the main teaching vehicle in the Spanish university system. With the advent of new methodological approaches, lectures are however losing their central role and student work is gaining weight, with classroom's hours increasingly devoted to personalised attention to the students [2]. In some cases, the new methodology of individual work by the students and in-class discussions have proven very productive. On the other hand, in courses with complicated contents or procedures, such as many Engineering courses, the lack of proper guided lectures setting the initial ground truth is a handicap that reflect on the students outcomes. In particular, the following problems have been detected among the students:

- Students need to work with a high number of materials from different sources, and they are usually forced to select, without prior knowledge, the best source for each particular content in the course.
- Copyright and intellectual property issues difficult the process of making the necessary materials available to students.
- Written materials are difficult to be accessed on libraries, where only a few copies are available.
- The individual work of the students increases the differences among them, complicating the teacher's work and the course preparation.
- Groups tend to be very populated and it is therefore difficult for the teacher to know how the students are progressing on an individual level.
- The work of the students does not have a supervision that accommodates their own rhythm.
- There is an excessive number of partial evaluations during the course. Students attention spans are often short and varying from one course to another as they only focus on the next exam.
- Students have difficulties on understanding which is the final level expected from their learning. This fact has also impact in following courses that rely on the current one.

In recent years, several academic institutions worlwide are proposing solutions to those problems making use of video-teaching tools [3]. Centres such as Stanford University, MIT or Georgia Tech have developed platforms to improve their learning curricula, contributing to their international visibility. This great scale distribution of contents allow that renowned professors can disseminate their knowledge to all those individuals interested in the subject. It is also relatively straightforward, by means of dubbing or the addition of subtitles, to offer a multi-language teaching experience using the aforementioned materials.

Since a few years ago, *e-learning* has consolidated as the most extended distant learning modality [4]. E-learning (also known as *on-line formation* or *virtual education*) consists on replacing the physical location of the learning centre by a computer. This modality is nowadays widely accepted and, as mentioned, included in the teaching portfolio of many relevant institutions. In traditional learning, the learner bases her progress on studying the materials and the personalised guide of a teacher who resolves the problems and complements the learning; *e-learning*, on the other hand, provides a generally bigger set of materials, permits self-pace and distant learning, but lacks the personalised guide of a tutor.

However, there is nothing to prevent the role of a supervisor —who resolves problems, corrects exercises and suggests new material— being incorporated into e-learning systems. The true reason behind most current systems not incorporating this aspect is the difficulty and complexity of developing and integrating such a feature, rather than in the pedagogical aspects. Some attempts have been made to simulate the presence of a teacher, creating a false appearance of personalisation. This is particularly interesting in *video-learning* courses [5], in which a teacher appears in the videos, simulating a real class.

E-learning tools that can be classified into two groups: of general purpose (such as *Moodle* or *WebCT* which allow the management of course and the creation of online learning communities), and the domain specific ones (which deal with particular areas of knowledge and includes higher levels of intelligent tutorization. Some examples are the ones teaching Nursery [6], Electric circuits [7], or air traffic controllers [8]).

The usual procedure of most of these tools consists in the tutor loading a set of materials in the platform, and the students visualising them at their convenience. Those e-learning systems treat all students in the same way, making no case of the differences among students. Therefore, they do not adapt themselves to the students, which is, on the other hand, a desirable feature.

The concept of *Intelligent Tutoring System (ITS)* [9] is defined as a system able to emulate the behaviour of a teacher in all relevant aspects [10], without the need of a real teacher and being the computer system the one in charge of guiding the students. One of the learning objectives of intelligent tutoring systems is that of adapting hypermedia courses to each final user [11] controlling the learning level and the contents navigability, adapting the available information and the methodology, explaining errors and suggesting solutions. In other words, an Intelligent Tutoring System is a system that prepares and shapes the learning material and procedures to the particular students and their performance.

The main objective of our research is to develop a video e-learning system integrating an Intelligent Tutoring System specially designed to the teaching of the undergraduate courses on *Artificial Intelligence* at the University of Granada. To do so, we propose an abstract model of Intelligent Tutoring System which will be later implemented and tested in a real environment (the aforementioned course, which has been taught by the authors for several years) to show its validity.

3 The Intelligent Tutoring System

3.1 Features

Our developed Intelligent Tutoring System presented the following aims:

- It models the knowledge state and abilities of the students.
- The student learning is driven in an adaptive way by means of constructing a personalised learning plan using the materials of the course.
- It teaches, examines and evaluates the students individually.
- It automatically evaluates the students answers by means of templates or teacher interaction.
- It controls the students progression level during the course.
- The educative material is mainly based on videos.

The Intelligent Tutoring System will therefore be able to anticipate if the student evaluation will be successful, shows how students respond to courses, and plans reinforcement activities. The teacher can also check the reinforcement plans and see the evolution of the students during the course.

One of the main features of an Intelligent Tutoring System is that of teaching each student in a personalised way [12,13]. The training methodology comprises the learning of each units concepts and the evaluation of the acquired knowledge. The tutor will supervise and analyse the students answers checking its correctness and potential failure trends.

If the students pass the exercises, she can move onto the next unit; if not, she is presented with new exercises and explanations related to her mistakes. Should the student keeps failing the unit, new reinforcement exercises and contents will be provided, and repeated, if needed. At the end of the course, the student will have to pass an exam covering all the course contents. This process is flexible enough to allow students to reinforce past units and review all contents.

As students are performing tests on a per unit basis, the tutor can assess on their progress, and the system can adapt the learning process. Tutors are presented with plenty of information about students performance and efforts. With the aim of avoiding the boredom of certain students, the system provides them with a sufficient degree of flexibility, as they can, if appropriate, choose the order of lessons, or skip them by directly attempting the tests.

3.2 Architecture

To achieve the expected behaviour for the Intelligent Tutoring System, several techniques from the *Knowledge Engineering* and *Artificial Intelligence* fields have been used [14]. In particular, intelligent systems are generally composed by several elements: a domain model, a student model, a teacher model, a pedagogical model. All of them are described in the following lines and depicted in Figure 1:

- Domain model: It provides the knowledge on the domain, providing the intelligence of the tutor. It is given as declarative (describing the facts) and procedural (the rules to execute a task) statements.
- Student model: It contains information about the student, such as: personal data, interactions and progress achieved.
- Teacher model: It contains the data related to the teacher.
- Pedagogical model: It is the main part of the Intelligent Tutoring System as it represents and manages the learning process and enables the functioning of the whole system through communication of all components. This model adapts the presentation of the material to each student using the information from the student's model, and creating a study plan (an ordered sequence of lessons). The pedagogical model comprises seven elements:
  - Evaluation model: It determines the student level and updates the interaction parameters in the student model.
  - Problem generation model: It generates the exercises and the units that the students need in their learning processes.
  - Problem resolution model: It deduces the right answer for a problem and prepares feedback on the students' answers.
  - Answer analysis model: It analyses the students' answers and check their correctness.
  - Unit revision generator model: It plans the lessons the students will have to review according to their learning progress.
  - Grade predicting model: It approximates the grades the students are likely to obtain at the end of the course.
  - Syllabus generation model: It suggests a course syllabus to the teacher.
- Graphical interface: This component is in charge of the interaction among the system and the users (student, teacher, administrator or visitor).

3.3 Implementation

Our Intelligent Tutoring System has been developed as a multi-agent system with a blackboard architecture (i.e. with several agents interchanging information in a common space) [15]. The agents are in charge of emulating the behaviour of a teacher using the aforementioned pedagogical model, and comprising, among others, the tasks of evaluating the questionnaires and the design of syllabus. The blackboard architecture enables a very modular system, in which each task can be implemented independently.

In particular, the system has a client/server architecture (see Figure 2), with the server distributing the web pages to the clients. The current implementation uses *Apache*, PHP, and *MySQL* (for the data repository). The graphical interface has been developed to be user-friendly.

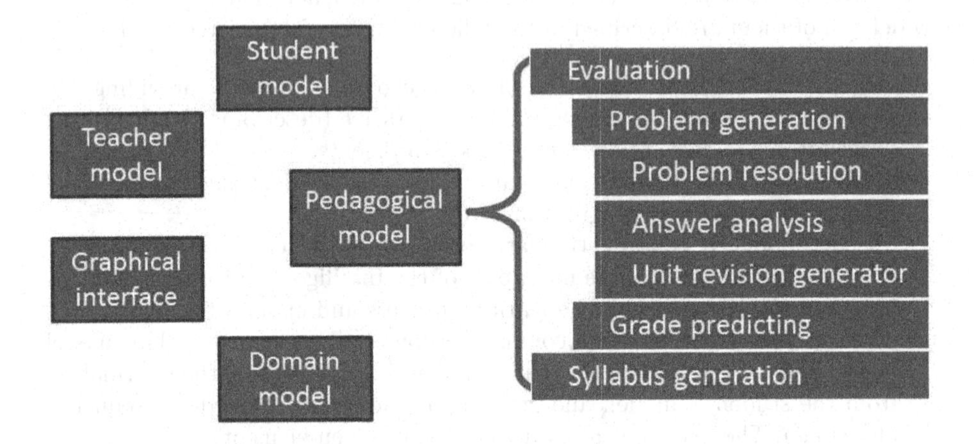

Fig. 1. Multi-agent architecture of the system

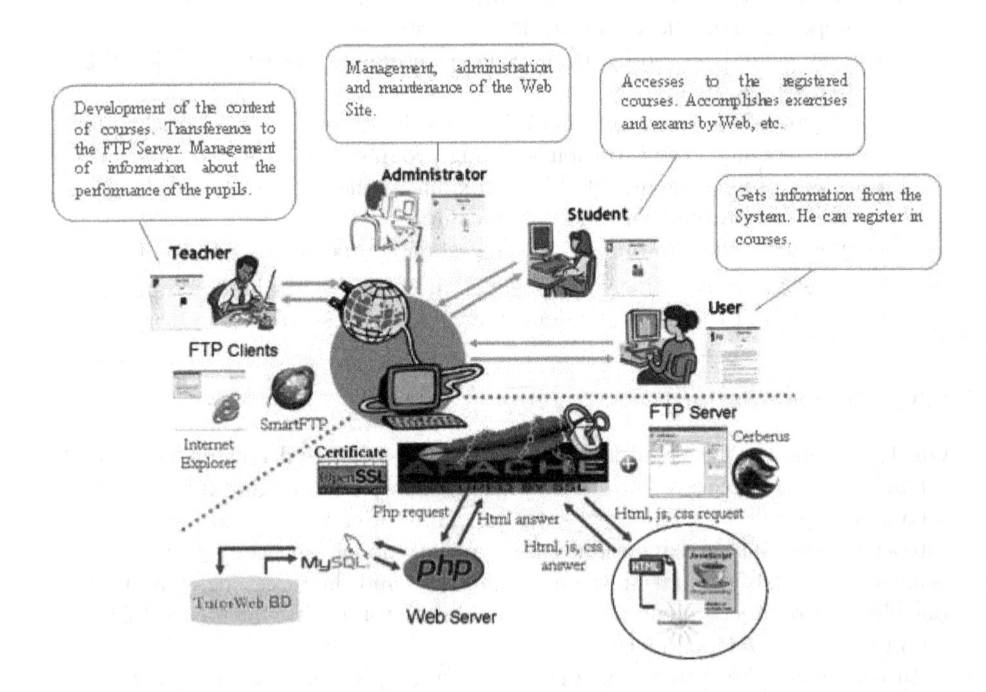

Fig. 2. Global view of the functionality of the system

The system leans on three basic components (that enables services to be added and removed dynamically, with no need to interrupt the system's execution):

- A basic common manager to both the client and the server. It takes upon the tasks of dynamic execution of code and intercommunication between processes in execution.
- A database that serves as common repository of the data driven by the system.
- Advanced services provided on top of the architecture for the end-users.

The system also offers additional services such as e-mail, forums, and video-conference, which allow all users to communicate between them.

3.4 Functionality

As depicted in Fig. 3, the devised tool has different user roles (administrators, teachers, students and unregistered users) with differentiated sets of privileges:

- Administrator: manages and maintains the web site, keeps the information about teachers and students, manages user accounts and courses, etc.
- Teacher: obtains information about existing courses and creates new ones. It can also visualise information about students and their progress and performance.
- Student: gathers information about teachers and courses, registers in new courses, and sees its progression and results.
- Visitor: any person without an account. It can get information about available courses and the teachers but cannot participate on them until getting registered.

4 Results

As previously stated, we aimed at facilitating the learning of the *Artificial Intelligence* course at *University of Granada*, by means of an e-learning system incorporating an Intelligent Tutoring System. The first step was the study of requirements of the system in order to achieve the objectives. This way, a set of features such as the potential number of concurrent students, the storage and connectivity needs, and the kind of interfaces to be developed. According to those requirements, an architecture to accommodate all of them was designed, proposing a central server storing all contents and the management software. All users will connect to it by means of a web system. On developing this Intelligent Tutoring System and its application, a prototype of e-learning server have been designed and developed, incorporating an ITS, for the multimedia teaching of a course. The system has incorporate the course's contents, achieving a sufficient degree of functionality.

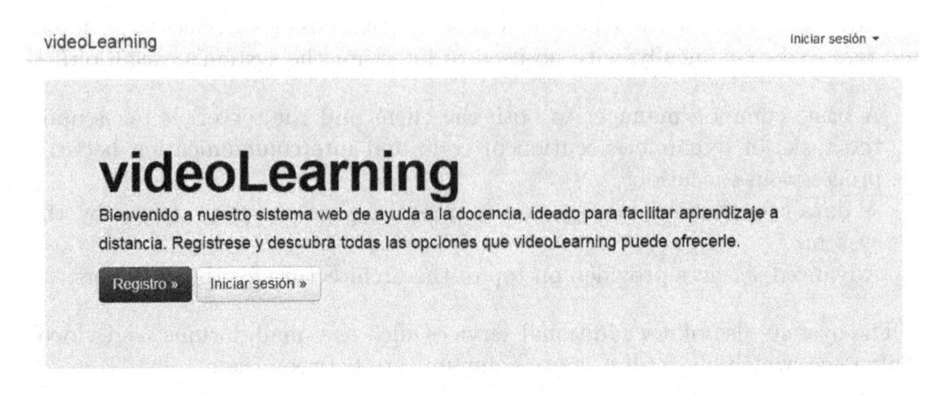

Fig. 3. Main screen of the web tool implementing the proposed Intelligent Tutoring System

4.1 Results on the Students' Learning

On the lights of the work done by the students with the platform, and the results they have gained in the final evaluation of the course, students have achieved a higher degree of comprehension of the course contents in comparison with previous years. According to the results of the evaluation of the course in the 2013/14 June call (the students which used the tool), up to 77.42% of the students attempted the evaluation, and from them, 91.67% passed the course. Those figures, in the first year in which the intelligent system was used, suggest the great educative potential of the developed tool. New analysis are to be made in the following years to confirm such expectations.

Besides that, students have referred better comprehension of the course contents thanks to the complementary materials and the system guidance and suggestions. Students also indicated the advantages of learning at their own pace, with the possibility of adapting their time and their efforts. In this way, some of the problems regarding an insufficient comprehension of the basic concepts in the course have been overcome. Guaranteeing the appropriate comprehension of these concepts was paramount for the success in the course, as many other concepts were building on top of the basic ones.

For teachers, the usage of the tool has allowed a deeper understanding of the student necessities, efforts and performance, making it possible to adapt the contents to particular groups of students. Additionally, course contents that were not successfully addressing the students' needs have been identified and will be remade in following revisions.

Finally, and due to positive outcomes of the experience, we are currently working on implementing the Intelligent Tutoring System in other related courses at Universidad de Granada.

5 Conclusions

This work described the development of an intelligent tutoring system that tries to imitates a real classroom on the learning procedures, and in particular, introducing a guiding entity that can personalised the learning process for each student. The developed Intelligent Tutoring System presents the following set of advantages:

- Flexibility: The system enables students to study at their own pace.
- Mobility: The physical presence in the classroom is no longer necessary.
- Effective: The outcomes of the learning process are evaluated.
- Cost reduction: Once developed, the system is easy to maintain.
- Wide coverage: It can be guaranteed that students will cover all the units and contents of the course.
- Personalisation: The contents and the learning process is adapted and personalised to the particular student.
- Easy updates: Course contents are easily updateable.

All in all, the Intelligent Tutoring System is a tool for managing, supporting and distributing learning courses in a web environment. It allows the creation and management, in a very flexible way, of platforms of e-learning that adapts the contents of the courses to each particular student, improving the outcomes of the learning experience.

References

1. Lau, R.W., Yen, N.Y., Li, F., Wah, B.: Recent development in multimedia e-learning technologies. World Wide Web 17(2), 189–198 (2014)
2. Avalos, A., Martín, L., Pérez-Urria, E., Pintos, B.: Problem-based learning and the use of new learning-teaching methodology. In: Procs. 4th Int. Technology, Education and Development Conference (INTED 2010), pp. 3004–3009 (2010)
3. Horton, W.: e-Learning by Design. Pfeiffer (2011)
4. Mayer, R., Clark, R.: e-learning and the Science of Instruction: proven guidelines for consumers and Designers of Multimedia Learning. Wiley and Sons (2011)
5. Lane, J., Fetherston, T.: Beyond U-tube: An innovative use of online digital video analysis in teacher education. In: Procs. 3rd International Conference on e-Learning. Academic Confereces Limited (2008)
6. Koutsojannis, C., Prentzas, J., Hatzilygeroudis, I.: A web based intelligent tutoring system teaching nursing students fundamental aspects of biomedical technology. In: Procs. 23rd Annual EMBS International Conference, pp. 4024–4027 (2001)

7. Yoshikawa, A., Shintani, M., Ohba, Y.: Intelligent tutoring system for electric circuit exercising. IEEE Transactions on Education 35(3), 222–225 (1992)
8. Kornecki, A., Hilburn, T., Diefenbach, T., Towhidnejad, M.: Intelligent tutoring issues for air traffic control training. IEEE Transactions on Control Systems Technology 1(3), 204–211 (1993)
9. Wenger, E.: Artificial Intelligence and Tutoring Systems. Morgan Kaufmann Publishers (1987)
10. Myller, N., Suhonen, J., Sutinen, E.: Using Data Mining for improving web-based course design. In: Procs. International Conference on Computers in Education, pp. 959–963 (2002)
11. Negnevitsky, M.: A knowledge based tutoring system for teaching fault analysis. IEEE Transactions on Power Systems 13(1) (1998)
12. Anderson, J., Boyle, C., Corbett, A., Lewis, M.: Cognitive modeling and intelligent tutoring. Artificial Intelligence 42, 7–49 (1990)
13. Hwang, G.: A concept map model for developing intelligent tutoring systems. Computers & Education 40(3), 217–235 (2003)
14. Schofield, J., Evans-Rhodes, D., Huber, B.: Artificial Intelligence in the classroom: The impact of a computer-based tutor on teachers and students. Social Science Computer Review 8(1), 24–41 (1990)
15. Wooldridge, M.: An Introduction to multiagent Systems. John Wiley & Sons (2002)

Part VII

Doctoral Consortium (DC)

Collision Avoidance System with Deceleration Control Applied to an Assistant Personal Robot

Eduard Clotet, Dani Martínez, Javier Moreno, Marcel Tresanchez, and Jordi Palacín

Department of Computer Science and Industrial Engineering,
University of Lleida, 25001 Lleida, Spain

Abstract. This paper presents the implementation of a dynamic collision avoidance system with deceleration control designed for Assistant Personal Robots based on the use of a LIDAR sensor.

Keywords: Collision avoidance, LIDAR sensor, acceleration control, omnidirectional wheels, remote controlled robot.

1 Introduction

Remotely operated robots can generate dangerous situations where the user is not able to control the robot (high latency, unstable connectivity…). LIDAR sensors are widely used to perform collision detection systems [1-2]. As an example, in [3] a robot designed to detect gas leaks use a LIDAR sensor in order to get information from the environment. In such robots, collision avoidance is performed by generating a predefined bounding box around the robot that must be big enough to avoid collisions at maximum robot velocity, but also small enough to be able to go through doors and narrow corridors.

2 Proposed Methodology

The hypothesis presented on this work is that it is possible to estimate the velocity of the mobile robot by using the information of the wheels and combine this information with a LIDAR sensor in order to dynamically define a collision bounding box according the velocity of the mobile robot. The mobile robot uses omnidirectional wheels (fig. 1) and has to prevent strong acceleration changes as they can cause the robot to flip over its horizontal axis.

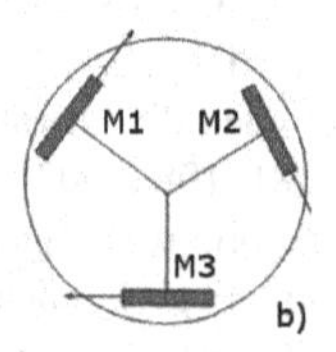

Fig. 1. a) Omnidirectional wheel. b) Omnidirectional wheels distribution.

J. Bajo et al. (eds.), *Trends in Prac. Appl. of Agents, Multi-Agent Sys. and Sustainability*,
Advances in Intelligent Systems and Computing 372, DOI: 10.1007/978-3-319-19629-9_25

3 Current Results

Figure 2 shows the results obtained when simulating the robot movement with: wheel diameter = 0.3 m, motor speeds of, $[M_1\ M_2\ M_3]$ = [20, -20, 0] rpm, robot velocity of 1m/s, frontal wall distance at 7m, and maximum negative acceleration of 0.7 m/s^2.

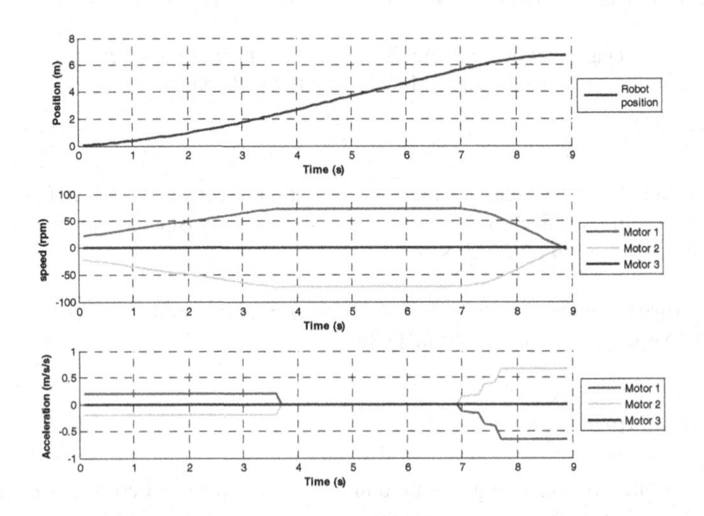

Fig. 2. Representation over time of the position, speed, and acceleration of the robot

4 Conclusions

The simulations performed have shown that it is possible to avoid drastic decelerations and collisions for forward trajectories. Future work will be focused in the general formulation of the procedure to include all possible mobile robot trajectories in order to avoid collisions and adequate mobile robot velocity to the environment.

Acknowledgements. This work is partially founded by Indra, the University of Lleida, and the RecerCaixa 2013 grant.

References

1. Ohki, T., Nagatani, K., Yoshida, K.: Collision Avoidance Method for Mobile Robot Considering Motion and Personal Spaces of Evacuees. In: International Conference on Intelligent Robots and Systems (2010 IEEE/RSJ), Taipei, Taiwan, pp. 1819–1824 (2010)
2. Surmann, H., Nütcher, A., Hertzberg, J.: An autonomous mobile robot with a 3D laser range finder for 3D exploration and digitalization of indoor environments. Robotics and Autonomous Systems 45(3-4), 181–198 (2003)
3. Martínez, D., et al.: A mobile robot agent for gas leak source detection. In: Bajo Perez, J., et al. (eds.) Trends in Practical Applications of Heterogeneous Multi-agent Systems. The PAAMS Collection. AISC, vol. 293, pp. 19–26. Springer, Heidelberg (2014)

An Agent-Based Decision Support Environment for Public Health

Viviane Maria Lelis Carvalho

Dpto. Lenguajes y Ciencias de la Computacion E.T.S.I. Informatica.
Universidad de Malaga, Bulevar Louis Pasteur,
35. Campus de Teatinos. 29071, Malaga, Spain
viviane@lcc.uma.es

Keywords: Decision support systems, public health, simulation.

1 Introduction

Decision support systems can be quite useful in public health management since it can provide new and relevant information regarding to population profile, incidence and behavior of diseases over time, as well as effectiveness of prevention methods. Such systems may also be useful if they are able to make predictions and trigger alerts about future events. Meningitis is an infectious disease with high mortality rate, especially in less devel-oped countries. DATASUS, the Informatics department of the Brazilian public health system (SUS), stores and makes available to professionals, researchers and public health managers a huge amount of health data, which is often underutilized. A retrospective study of these data to chart a disease behavior model in a particular region and the simulation of possible scenarios could be quite helpful in making decisions to reduce the incidence and mortality of this disease. Multi-Agent Systems (MAS) are highly appropriate to the complex environment of health. More specially Agent-Based Modeling and Simulation (ABMS) has been used to support decision making in different scenarios related to health [3]. For example, in emergency departments, for predicting the effects of patient's derivation policies [4]; in the area of computational epidemiology, evaluating the spread of infectious diseases [2]; or for validating the model for supporting emergency evacuation decision making [5]. Tree-based decision models are decision-making components of medical curriculum. These models can supplement the clinical practice guidelines or even they can be used as a standalone decision tool. For example, in [1] a tree based decision model was developed to predict the severity of pediatric asthma exacerbations.

2 Proposal and Preliminary Results

Our goal consists in exploring the use of MAS technology to develop tools that will assist policy makers in public health management, making use of ABMS and

J. Bajo et al. (eds.), *Trends in Prac. Appl. of Agents, Multi-Agent Sys. and Sustainability*,
Advances in Intelligent Systems and Computing 372, DOI: 10.1007/978-3-319-19629-9_26

data mining techniques. We have data of meningitis cases (happened in Bahia, Brazil, during the last 10 years), provided by DATASUS. With those data we are exploring two working line: First we are constructing an agent-based simulator. For this purpose, we have identified the different types of agents involved in the treatment of meningitis. Agent behavior is being constructed following a data-driven approach. Accordingly, a state machine would describe each behavior where transitions between states are determined according to probabilities obtained from our dataset. Our simulator may help the policy makers to draw a profile of the cure and mortality during the analyzed period, 10 years, and to identify changes in its behavior that may have occurred over these years. NetLogo [7], a high level platform particularly well suited for modeling complex systems developing over time, is being used to construct our simulator. The second research line we are exploring consists in studying the dataset in order to obtain statistical evidence suggesting the effectiveness of prevention vaccines, the principal agents of contamination, the seasonality of the disease and to evaluate the most suitable strategies for an early intervention that would reduce the mortality rate of meningitis. Our first results of that analysis suggest the existence of a gap between the data and recommendations for the diagnosis and management of bacterial meningitis. Those recommendations are collected in the Clinical Practice Guidelines (CPG) [6] commonly used by medical doctors. The found gap is due to the lack of resources in less developed regions where some useful tests for a more accurate diagnosis cannot be made. As a consequence, our goal is to apply data mining techniques to obtain a tree decision that would assist in the diagnosis of meningitis and could lead to elaborate an adaption of the CPG to the particular characteristics of rural areas de Brazil.

References

1. Farion, K., et al.: A tree-based decision model to support prediction of the severity of asthma exacerbations in children. J. Med. Syst. 34, 551–562 (2010)
2. Linard, C., Poncon, N., Fontenille, D., Lambin, E.F.: A multi-agent simulation to assess the risk of malaria re-emergence in southern France. Ecollogical Modelling 220, 160–174 (2009)
3. Macal, C.M., North, J.M.: Tutorial on agent-based modeling and simulation. Journal of Simulation 4, 151–152 (2010)
4. Taboada, M., et al.: Using an agent-based simulation for predicting the effects of patients derivation policies in emergency departments. In: 2013 International Conference on Computational Science (2013)
5. Tian, Y., Zhou, T.Z., Yao, Q., Zhang, M.: Use of an agent-based simulation model to evaluate a mobile-based system for supporting emergency evacuation decision making. Journal of Medical Systems, 38–149 (2014)
6. Tunkel, A.R., et al.: Practice guidelines for the management of bacterial meningitis. Clinical Infectious Diseases 39, 1269–1284 (2004)
7. Wilensky, U.: Netlogo. Center for Connected Learning and Computer-Based Modeling, Northwestern University Evanston (1999)

Selected Methods of Model Checking Using SAT and SMT-Solvers*

Agnieszka M. Zbrzezny

IMCS, Jan Długosz University,
Al. Armii Krajowej 13/15, 42-200 Częstochowa, Poland
agnieszka.zbrzezny@ajd.czest.pl

Abstract. The objectives of this research are to further investigate the foundations for novel SMT and SAT-based bounded model checking (BMC) algorithms for real-time and multi-agent systems. A major part of the research will involve the development of SMT-based BMC methods for different kinds of Kripke structures, interpreted systems for different kinds of temporal languages.

Keywords: Bounded model checking, SMT, SAT, Temporal Logic, PhD thesis extended abstract.

1 Introduction

Model checking [1] is as a method for automatic and algorithmic verification of finite state concurrent systems.Automated verification of real-time systems (RTS) and multi-agent systems (MAS), performed by the analysis of their models is a very important subject of research. This is highly motivated by an increasing demand to verify safety critical systems, i.e., time-dependent systems, failure of which could cause dramatic consequences for both people and hardware. In view of this, there is an obvious need to develop efficient SMT/SAT-based verification methods which could be used in practice.

2 SAT and SMT

SAT-based BMC [2] uses a reduction of the problem of truth of a modal formula in a model (transition system) to the problem of satisfiability of formulae of the classical propositional calculus, i.e. SAT-problem. The reduction is achieved by a translation of the transition relation and a translation of a given property to formulae of classical propositional calculus. It should be emphasised that for a given temporal logic, BMC is mainly used to disprove safety properties and to prove liveness properties.

The SMT problem [3] is a generalisation of the SAT problem, where Boolean variables are replaced by predicates from various background theories. SMT

* The study is co-funded by the European Union, European Social Fund. Project PO KL "Information technologies: Research and their interdisciplinary applications", Agreement UDA-POKL.04.01.01-00-051/10-00.

J. Bajo et al. (eds.), *Trends in Prac. Appl. of Agents, Multi-Agent Sys. and Sustainability*, Advances in Intelligent Systems and Computing 372, DOI: 10.1007/978-3-319-19629-9_27

generalises SAT by adding equality reasoning, arithmetic, fixed-size bit-vectors, arrays, quantifiers, and other useful first-order theories. Using SMT to express different problems has important advantages over SAT. If one uses SAT, then, for example, data must be encoded into a Boolean representation: a bit-vector must be represented as just its individual bits. In contrast, an SMT encoding can represent the bit-vector directly, and may be able to reason more efficiently at the bit-vector level of abstraction, without resorting to bit-level reasoning.

3 Methodology

The main aim is to compare the existing SAT-based BMC algorithms with our new SMT-based BMC techniques for the same models.

Implementations. The implementations are written in C++.

Scenarios and Benchmarks. In the area of formal verification it is customary to use scalable benchmarks in order to demonstrate the effectiveness of verification tools.

4 Conclusions

We proposed, implemented, and experimentally evaluated some SAT-based [4] and SMT-based BMC approaches i.e. for WECTLK interpreted over the weighted interpreted systems (extended abstract accepted for AAMAS 2015), for RTECTL properties interpreted over the simply-time systems (accepted for DCAI 2015) and for ECTL* [5] properties. We have compared our SMT-based BMC methods with the corresponding SAT-based BMC methods. The experimental results show that the approaches are complementary. Also an observation of experimental results leads to the conclusion that the SAT-based BMC for uses less memory comparing to the SMT-based BMC. This is a novel and interesting result, which shows that the choice of the BMC method should depend on the considered system.

References

1. Baier, C., Katoen, J.-P.: Principles of Model Checking. MIT Press (2008)
2. Biere, A., Cimatti, A., Clarke, E., Fujita, M., Zhu, Y.: Symbolic model checking using SAT procedures instead of BDDs. In: Proceedings of the ACM DAC 1999, pp. 317–320 (1999)
3. Biere, A., Heule, M., van Maaren, H., Walsh, T. (eds.): Handbook of Satisfiability. Frontiers in Artificial Intelligence and Applications, vol. 185. IOS Press (2009)
4. Woźna-Szcześniak, B., Szcześniak, I., Zbrzezny, A.M., Zbrzezny, A.: Bounded model checking for weighted interpreted systems and for flat weighted epistemic computation tree logic. In: Dam, H.K., Pitt, J., Xu, Y., Governatori, G., Ito, T. (eds.) PRIMA 2014. LNCS (LNAI), vol. 8861, pp. 107–115. Springer, Heidelberg (2014)
5. Zbrzezny, A.M., Zbrzezny, A.: A comparison of SAT-based and SMT-based bounded model checking methods for ECTL*. In: Proceedings of the Workshop CS&P 2014. CEUR Workshop Proceedings, vol. 1269, pp. 293–300 (2014)

Exploration/Exploitation in Stochastic Distributed Constraint Optimization Settings

Julius Pfrommer*

Fraunhofer Institute of Optronics,
System Technologies and Image Exploitation,
Karlsruhe, Germany
julius.pfrommer@iosb.fraunhofer.de

Abstract. In recent years, important results have been achived in decision-making in uncertain environments, where actions have a direct reward as well as long-term ramifications by bringing in additional information used to improve future decisions. We propose a line of work where this exploration/exploitation tradeoff is applied to distributed settings with interacting independent agents.

1 Introduction

Many agent-based applications need to interact with an uncertain environemt. Whenever the environment state or even the dynamics of the interaction change, the system needs to adapt. The tradeoff between (a) keeping the state information and system model up-to-date and (b) to maximize short-term utility based on current knowledge is nontrivial. Optimizaton in such settings is especially difficult when actions need to be coordinated between independent agents.

2 Relevant Fields of Research

Decentralized Constraint Optimization The Max-Plus algorithm, an instance of the Generalized Distributive Law family [6], is known to solve the Distributed Constraint Optimization Problem (DCOP) of maximizing the summed utility functions of interacting agents [4]. The graph-structure that encodes the utility inter-dependencies between agents is also used for a range of additional applications relevant to this case, such as probabilistic [5] and causal [9] reasoning.

Exploration/Exploitation. Auer et al. [1] initially sought to minimize the *regret* of actions in multi-armed bandit settings via the notion of an Upper Confidence Bound (UCB). Since then, UCB-based techniques have been applied to solve Markov Decision Problems (MDP) [3] and Partially Observable Markov Decision Problems (POMDP) [10]. Some efforts were made to apply UCB to solving

* The work of Julius Pfrommer is partially supported by the BMBF (German Ministry of Education, Science, Research and Technology) under grant number 01IS12053.

J. Bajo et al. (eds.), *Trends in Prac. Appl. of Agents, Multi-Agent Sys. and Sustainability*, Advances in Intelligent Systems and Computing 372, DOI: 10.1007/978-3-319-19629-9_28

general DCOP problems [8]. Recent of work has begun to apply DCOP in online learning settings in stochastic environments based on reinforcement learning [7] and game-theoretic extensions, based on the notion of potential games [2]. However, the authors mainly envisage decentralized sensing settings and do not deal with acting agents, where the interaction between sensing, actuation and learning further increases the optimization complexity.

3 Background and Research Goal

Recently, we developed a generalization of the well-known Max-Plus algorithm, that can be used to exactly solve DCOP problems also in settings with cyclic utility inter-dependencies between agents (publication pending, code and examples are available under `https://github.com/jpfr/pygmalion`). On the application side, we develop a decentralized industrial control system as part of a BMBF-funded project. It is intended to capture additional stochastic runtime information that cannot be determined up front, such as energy consumption and process yield. Our aim is to adapt our variation of Max-Plus to decentrally optimize the overall system behavior in order to react to the learned system dynamics and anticipated future learning opportunities.

References

1. Auer, P., Cesa-Bianchi, N., Fischer, P.: Finite-time analysis of the multiarmed bandit problem. Machine Learning 47(2-3), 235–256 (2002)
2. Chapman, A.C., Leslie, D.S., Rogers, A., Jennings, N.R.: Learning in unknown reward games: application to sensor networks. The Computer Journal, bxt082 (2013)
3. Kocsis, L., Szepesvári, C.: Bandit based monte-carlo planning. In: Fürnkranz, J., Scheffer, T., Spiliopoulou, M. (eds.) ECML 2006. LNCS (LNAI), vol. 4212, pp. 282–293. Springer, Heidelberg (2006)
4. Kok, J.R., Vlassis, N.: Using the max-plus algorithm for multiagent decision making in coordination graphs. In: Bredenfeld, A., Jacoff, A., Noda, I., Takahashi, Y. (eds.) RoboCup 2005. LNCS (LNAI), vol. 4020, pp. 1–12. Springer, Heidelberg (2006)
5. Koller, D., Friedman, N.: Probabilistic graphical models: principles and techniques. MIT press (2009)
6. Kschischang, F.R., Frey, B.J., Loeliger, H.A.: Factor graphs and the sum-product algorithm. IEEE Transactions on Information Theory 47(2), 498–519 (2001)
7. Nguyen, D.T., Yeoh, W., Lau, H.C., Zilberstein, S., Zhang, C.: Decentralized multi-agent reinforcement learning in average-reward dynamic dcops. In: Proceedings of the 2014 International Conference on Autonomous Agents and Multi-agent Systems, pp. 1341–1342. International Foundation for Autonomous Agents and Multi-agent Systems (2014)
8. Ottens, B., Dimitrakakis, C., Faltings, B.: Duct: An upper confidence bound approach to distributed constraint optimization problems. In: Proceedings of the National Conference on Artificial Intelligence, pp. 528–534 (2012)
9. Pearl, J.: Causality: models, reasoning, and inference. Cambridge University Press (2000)
10. Silver, D., Veness, J.: Monte-carlo planning in large POMDPs. In: Advances in Neural Information Processing Systems, pp. 2164–2172 (2010)

Human-in-the-Loop Multi-agent Approach for Airport Taxiing Operations

François Lancelot[1,2,3], Mickaël Causse[2],
Nicolas Schneider[1], and Marcel Mongeau[3]

[1] Airbus Group Innovations, Toulouse, France
[2] ISAE, Université de Toulouse, Toulouse, France
[3] ENAC, Toulouse, France

Abstract. This paper describes a study to develop an intelligent air traffic ground control operation system that is able to integrate initiatives from both controllers and autonomous vehicles. We use a multi-agent approach to optimize dynamically a routing and scheduling task while taking into account real-time human suggestions. We expect that such mixed-initiative systems combined with autonomous vehicles will permit reduction of congestion and fuel consumption in large airports.

Keywords: multi-agent, human-in-the-loop, taxiing, optimization.

Air traffic control (ATC) ground operations consist of real-time, safe, and efficient management of aircraft and vehicles on taxiways and runways. Ground operations are characterized by unexpected events (e.g., infrastructure and aircraft maintenance checks) which can require real-time reactions, making it difficult to develop optimal and automatic schedules. It is not surprising that airports have increasingly become a bottleneck within the air transportation network. Reducing ground congestion by optimizing vehicle trajectories is necessary. The ultimate aim of this study is not only to reduce delays, but also to improve safety and to reduce fuel consumption.

While centralized approaches like mixed-integer linear programming (e.g., [1]) and genetic algorithm (e.g., [2]) have successfully solved such ground movement problems, they are known for single-point failures and do not integrate human initiative. Human initiative remains mandatory as the system needs human expertise and knowledge in many situations. For instance, the system could be unaware of an unexpected taxiway closure and continue sending aircraft on this taxiway. Multi-Agent Systems (MAS) are capable of supporting large scale dynamic environments and can integrate the constraints and objectives of the operators by modeling agents starting from the natural description of the problem. However, most research in MAS has focused on developing fully automatic algorithms [3] [4].

The objective of this work is to build a decentralized multi-agent system that can suggest routing solutions via a tactile interface to optimize dynamically a routing and scheduling task while taking into account human suggestions in real

J. Bajo et al. (eds.), *Trends in Prac. Appl. of Agents, Multi-Agent Sys. and Sustainability*, Advances in Intelligent Systems and Computing 372, DOI: 10.1007/978-3-319-19629-9_29

time. This system is to be used in tandem with autonomous vehicles able to tow an aircraft with its jet engines off.

Our contribution is threefold. First, a MAS has been developed for optimization. This work is based on the delegate multi-agent pattern [5]. Combining lightweight agents and heavyweight agents offers flexibility and scalability. A reservation-based control mechanism [6] has been established for managing the resources. In addition, resources adapt themselves to give priority to agents that have difficulty reaching their goal.

Second, the MAS is integrated into the ATC ground operations problem and is incorporating standard taxiway traffic flow and separation constraints. Each taxiway is managed by a reservation planner, which estimates the duration of an operation for a given arrival time, and vehicle speed using a constraint satisfaction solver. Each vehicle dynamically updates its criticality (computed using real waiting and taxiing times) and evaluates each possible trajectory by sending exploration ants. The vehicles that have the ability to adjust their speeds build a new feasible trajectory so as to avoid waiting time.

Third, the MAS dynamically takes into account human initiative. Operators can suggest or impose new paths but the vehicle agents will continue to explore other possibilities in order to be able to propose a better suggestion to the controller at any time. Users can also designate priority usage between vehicles for the taxiways. An experience database logs the user choices, and the MAS uses it to adapt its next suggestions.

This system is currently being tested on a small environment involving intersections and detaching stations. It will then be evaluated in the MoTa simulation platform [7] (Roissy Charles de Gaulle Airport, Paris, France).

References

1. Roling, P.C., Visser, H.G.: Optimal airport surface traffic planning using mixed-integer linear programming. International Journal of Aerospace Engineering 2008(1), 1–11 (2008)
2. Gotteland, J.-B., Durand, N.: Genetic algorithms applied to airport ground traffic optimization. In: Proc. of the Congress on Evolutionary Computation, Canberra (2003)
3. You, J., Han, S.-C.: Taxi route optimization algorithm of airport surface based on multi-agent. Journal of Traffic and Transportation Engineering 9, 109–112 (2009)
4. Claes, R., Holvoet, T., Weyns, D.: A decentralized approach for anticipatory vehicle routing using delegate multiagent systems. IEEE Transactions on Intelligent Transportation Systems 12(2), 364–373 (2011)
5. Holvoet, T., Weyns, D., Valckenaers, P.: Patterns of delegate MAS. In: 3rd IEEE Int. Conference on Self-Adaptive and Self-Organizing Systems, San Francisco (2009)
6. Dresner, K., Stone, P.: Multiagent traffic management: A reservation-based intersection control mechanism. In: 3rd Int. Joint Conference on Autonomous Agents and Multiagent Systems, New York, pp. 530–537 (2004)
7. Chua, Z., Cousy, M., Andre, F., Causse, M.: Simulating air traffic control ground operations: Preliminary results from project modern taxiing. In: The 4th SESAR Innovation Days, Madrid (2014)

Trait Emotional Intelligence: Modelling Individual Emotional Differences in Agent-Based Models

Andrea Scalco

University of Verona, Department of Philosophy,
Education and Psychology, Verona, Italy
andrea.scalco@univr.it

1 Problem Statement

From a psychological perspective, the construct of emotional intelligence (EI) is strictly related to the individual experience of emotions. Despite the several prominent theories emerged during the years [1,2,3], all authors recognize that emotional intelligence concerns two main aspects of individual emotional experience: the faculty to recognize and control own emotions, and the ability to understand and influence what others feel. In this way, EI can be considered as a superior dimension useful to the perception, control and management of emotions: from an individual viewpoint (*i.e.* what particular emotion am I feeling in this moment?), as well as for the social dimension (*i.e.* what are others feeling right now?). In this way, the aim of the Ph.D. project is the development of agent-based models, especially in order to recreate the social mechanisms studied by social and organizational psychology (*e.g.* teamwork; competition and negotiation), able to take into account the individual differences related to the management of the emotional experiences and affecting decision-making processes.

2 Hypothesis

We believe that the implementation of trait emotional intelligence inside agents endowed with beliefs, desires and intentions (BDI) can improve the capability of agent-based models to reproduce individual differences related with the emotional sphere. In this way, we suppose that simulations of social processes can unearth the dysfunctions that occasionally are observed in real world within organizations (e.g. trigger off emotional distress to push out an individual perceived as a rival from an organization).

3 Related Work

An extensive book about how to model emotional intelligence has been published by Chakraborty and Konar [4]. The content follows the model of EI proposed by [1]: that is, a competency based approach. But, from a psychological perspective, when EI is conceived as a competence (*i.e.* ability) it does not take into account the investigation of individual differences. Contrarily, it is possible to understand individual differences

J. Bajo et al. (eds.), *Trends in Prac. Appl. of Agents, Multi-Agent Sys. and Sustainability*,
Advances in Intelligent Systems and Computing 372, DOI: 10.1007/978-3-319-19629-9_30

among people when the EI is conceptualized as a stable trait of personality: coherently, the present project follows the *trait emotional intelligence theory* as proposed by Petrides [trait EI; 2]. More importantly, if we assume the competency approach, we have to assume that EI is always adaptive. On the contrary, EI is not always adaptive, as confirmed by [5, p. 1356], "its adaptive value will vary depending on the context".

4 Proposal

We propose to implement the concept of trait EI associating it with the agents' decision-making processes in order to improve the complexity related to the individual differences: that is to say, modelling individual *emotional* differences. The proposal rely on BDI paradigm and try to connect it with trait emotional intelligence: precisely, the present proposal does not want to focus on emotions as themselves, instead, it propose to model, what we can call, the "emotional management" part of an agent through the introduction of trait EI into BDI paradigm. Whereas virtual emotions affect the beliefs of a virtual intelligent agent, the EI affect primarily agent's desires and intentions. Indeed, a BDI agent must follow different paths of action not only on the base of their affect ("states of mind"), but depending on several conditions related to management of individual emotional experience, that is, its trait EI. For instance: (*i*) how the agent is good to recognize its own feeling? Could intervene some errors? In addition (*ii*) how the agent is skilled to predict the emotional consequences of its action in relation to others agents? We think that, from a modelling perspective, the first question could, for example, directly influences the agents' "state of mind", whereas the second one could determine its interaction ability defining in this way its action plan.

5 Reflections

As computational models are considered able to solve out contradictory theoretical issues [6], the hope is even to advance the knowledge regarding this psychological construct. EI is actually a highly discussed construct: therefore, we believe that psychological research could benefit from the methods provided by computer science in order to get closer to solve out the theoretical issues concerning emotional intelligence.

References

1. Salovey, P., Mayer, J.D.: Emotional Intelligence. Imagin. Cogn. Pers. 9, 185–211 (1990)
2. Petrides, K.V.: Trait Emotional Intelligence Theory. Ind. Organ. Psychol. 3, 136–139 (2010)
3. Bar-On, R.: The Bar-On model of emotional intelligence: A valid, robust and applicable EI model. Organ. People 14, 27–34 (2007)
4. Chakraborty, A., Konar, A.: Emotional Intelligence A Cybernetic Approach. Springer, Berlin (2009)
5. Sevdalis, N., Petrides, K.V., Harvey, N.: Trait emotional intelligence and decision-related emotions. Pers. Individ. Dif. 42, 1347–1358 (2007)
6. Vancouver, J.B., Scherbaum, C.A.: Do we self-regulate actions or perceptions? A test of two computational models. Comput. Math. Organ. Theory 14, 1–22 (2008)

Intelligent Classifier for E-Nose Systems

Dechen Pelki[1], Javier Bajo[1], and Sigeru Omatu[2]

[1] Department of Artificial Intelligence, Technical University of Madrid, Madrid, Spain
[2] Osaka Institute of Technology, Osaka, Japan
pelki.dechen.pelki@alumnos.upm.es, jbajo@fi.upm.es,
omatu@rsh.oit.ac.jp

Abstract. This paper briefly discusses the state-of-the-art of e-noses and classifiers used in analyzing the response data from E-Nose systems and presents an idea about how to face off this kind of problems using ensembles of classifiers.

Keywords: E-Nose system, ensemble of classifiers.

1 Introduction

An E-Nose is an instrument used for the automated detection and classification of odors, vapors and gases, thus mimicking the human olfactory apparatus [3]. It is used in the perfumes, food, beverages and biomedical industries to classify various complex odors. A general E-Nose system consists of a sample handling, a detection and a data processing system. After considering the samples and methods specifications for the sample handling, the detection system consists of an array of sensors and then the pattern recognition techniques for analyzing the response data generated by the detectors [5]. In this paper, we revise the state-of-the-art about e-noses and the classifiers used in the existing e-noses. Then we obtain some preliminary conclusions and propose an ensemble of classifiers as an innovative technique for an e-nose system developed in Japan.

2 Related Work

Most of the applications for e-noses concentrate on four major areas; food, medical diagnosis, environmental monitoring and bio-process control [7]. Rodriguez *et al.* [6] used an A-Nose to discriminate Colombian coffee into simple and complex odors using PCA and MLP BPNN with LOO cross validation method. A portable e-nose Pen2 identified the quality grade of green tea by extracting feature vectors using PCA and validating using LDA and BPNN [10]. It is possible to directly classify the sensor data using BPNN only; using a TGS 800 series Smart E-Nose for controlling coffee quality [8] but it leads to overfitting. It is also possible to extract features to reduce data dimension using PCA only without validating or using any supervised classifiers such as the work by Berna *et al.* [2], in comparing 2 e-noses for detecting changes in tomato aroma profiles of two different cultivars and a quartz crystal microbalance based e-nose for detection of bacterial contaminated milk [1]. However, Dutta *et al.*

J. Bajo et al. (eds.), *Trends in Prac. Appl. of Agents, Multi-Agent Sys. and Sustainability*,
Advances in Intelligent Systems and Computing 372, DOI: 10.1007/978-3-319-19629-9_31

[4] combined 3D-scatter plot, Fuzzy C Means, SOM for extracting features, and used MLP, PNN and RBF for classifying the 6 eye bacteria data. This proposal enhanced the performance of Cyranose 320 e-nose, but it can be very difficult to implement and incorporates a high degree of complexity. Chen *et al.* [3] employed a MDC with a combination of KNN, LDA and PNN, which provided an increased overall classification accuracy compared to the use of any single individual classifier.

3 Conclusions and Research Goal

E-Noses mostly use feature extraction methods followed by a pattern recognition method. The typical feature extraction linear methods used are PCA, LDA and ICA [9]. The task of a classifier is to use the feature vector provided by the feature extractor to assign the object it represents to a category [7]. Currently, the existing E-Nose systems employ a feature extraction method followed by a single classifier for analyzing the response data except for two cases where the feature extraction methods have been combined and an appropriate classifier that performs specifically better with the combination has been suggested in [3][4]. The methods proposed are also specific to the kind of odor that is being detected. Our aim is to propose an ensemble of classifiers to analyze the response data, trying to take advantage of the combined responses of classifiers and focusing on a particular e-nose system created at the OIT in Japan. More specifically, we propose the use of MLP and SVM to explore the combination of different classifiers in an effective manner.

References

1. Ali, Z., O'Hare, W.T., Theaker, B.J.: Detection of Bacterial Contaminated Milk by means of a Quartz crystal Microbalance Based Electronic Nose. Journal of Thermal Analysis and Calorimetry 71, 155–161 (2003)
2. Berna, A.Z., Lammertyn, J., Saevels, S., Natale, C.D., Nicolai, B.M.: Electronic nose systems to study shelf life and cultivar effect on tomato aroma profile. Sensors and Actuators B 97, 324–333 (2004)
3. Chen, H., Goubran, R.A., Mussivand, T.: Improving the Classification Accuracy in Electronic Noses Using Multi-Dimensional Combining (MDC). IEEE (2004)
4. Dutta, R., Hines, E.L., Gardner, J.W., Boilot, P.: Bacteria classification using Cyranose 320 electronic nose. BioMedical Engineering Online (2002)
5. Peris, M., Escuder-Gilabert, L.: A 21st century technique for food control: Electronic noses. Elsevier (2009)
6. Rodriguez, J., Duran, C., Reyes, A.: Electronic Nose for Quality Control of Colombian Coffee through the Detection of Defects in "Cup Tests". Sensors (2010)
7. Scott, S.M., James, D., Ali, Z.: Data Analysis for electronic nose systems. Springer (2006)
8. Shilbayeh, N.F., Iskandarani, M.Z.: Quality Control of Coffee Using an Electronic Nose System. American Journal of Applied Sciences 1(2), 129–135 (2004)
9. Stone, J.V.: Independent component analysis: an introduction. Trends Cogn. Sci. 6, 59 (2002)
10. Yu, H., Wang, J., Xiao, H., Liu, M.: Quality grade identification of green tea using the eigenvalues of PCA based on the E-nose signals. Sensors and Actuators B 140, 378–382 (2009)

Author Index

Printed in the United States
By Bookmasters